The Creative Principle

The Creative Principle

A Cosmology for the 21st Century

William O. Joseph

ISISS Inc.

Somers, Montana

ISISS Inc.
P.O. Box 188
Somers, Montana 59932
Isiss@centurytel.net

Editing by Stephanie Gunning
Interior design by Gus Yoo
Cover design by A.J. Joseph

ISBN:9780615682037

1. Science and religion 2. Cosmology 3. Evolution 4. Spirituality
5. Law of attraction 6. Law of unintended consequences
7. The quickening 8. Yugas 9. Philosophy.

For Arie and Matt, Amar and Leslie, Debi and Lionel

Contents

Acknowledgments

First and foremost, this book is the result of a 25-year conversation with my wife, Marshelle, for which nothing in writing could ever adequately convey my appreciation. Second, many thanks to my friends and colleagues Michelle Skaletsky-Boyd, Peter Lengsfelder, Linda Graf, Daniel Giamario, Cayelin Castell, Lois Katz, Clif Palmer, and Paul Tribble. Finally, to my editor Stephanie Gunning, who demands that I think deeply about what I am saying and whose insights I could not do without: Thank you Stephanie, I'm looking forward to the next one.

Acknowledgments

Introduction

The following information flows from an imperfect understanding. Like most of the people who will be attracted to this book, I am simply a student of reality. I present the understanding described in these pages from the point of view of a country boy. Most of what I have learned in life has come to me experientially. Everything here, with the exception of what appears in quotes, is an opinion based upon that experience. I'm not attempting to prove what I am saying to you. I'm attempting to provide a dialogue that is engaging enough for you to be motivated to prove what is being said through your own experience.

A lot of what I've learned in my studies has come from the long-term relationship I've enjoyed with the natural world. In my early 20s, I spent several years living primitively in the mountains of Colorado. I explored the Rocky Mountains on horseback over a 12-year period in my 40s and 50s. Through these experiences I came to understand that the ancient secrets of creation are encoded in nature: They are there to see if you know how to look.

I want to describe a way to look at our role in creation that's both new and ancient at the same time. The story I'm going to tell you blends myth with history, science, religion, and philosophy. It's a complex story if you are unfamiliar with some of the concepts contained in it. In order to tell it I'll have to fill in those concepts one by one so that we can begin to build a common vocabulary. That vocabulary will be needed to give meaning to what I call the "Creative Principle." After you've lived with this story for a while it turns into a very simple tale about a relationship we all have with the Creative Principle. My intent in telling this story is to create a new sense of why we are here and what there is to do about the state the world is in.

I hope to show you that the experience we refer to as "life" is generated from within us, and that all of our efforts to fix the symptoms of our worldwide distress won't produce the outcomes we seek if we don't change our inner world first. In order for each of us to make appropriate changes to our own inner world, it would be helpful to know why things have evolved as they have on our planet.

We've been telling ourselves a story about "how it really is" in the universe. It's a story that's out of date relative to what is actually happening. I want to describe to you how this story is part of the root cause of everything humanity has manifested. I will admit up front that the new story I'm going to tell you is about a belief system.

When does it become a priority to re-examine the belief systems that have formed the religious, philosophical, or political foundation of a people or a nation for centuries? When old belief systems founder, placing the well-being and even the survival of large populations at stake, we should stop, re-evaluate what we believe in, and make corrections in the belief systems that manifest our reality.

Most folks have political beliefs, religious beliefs, sociological beliefs, scientific beliefs, and so on. Since the pace of modern life demands that we react quickly to what goes on around us, we group those beliefs into belief systems for easy access. Try as we might to react spontaneously to every situation that arises in life, it's very easy to revert to the default setting of our belief systems. We are unable to spend the time necessary to create a fresh response to each stimulus, carefully examining ourselves to make sure that everything we are about to say or do embodies the highest principles of which we are aware. So we consult the "playbook" and run a pre-recorded message stored in the archives of our personal belief systems. It is the clash of personal belief systems that is causing so much emotional strife in our world.

Our world is made up of dynamic polarities: hot/cold, big/little, fast/slow. We live in a world of duality. If the clash of belief systems exists and causes strife, then by the principle of duality a belief system must exist that has the potential to unify all of humanity, and in which strife is greatly reduced or eventually eliminated. I'm going to propose such a belief system to you. It is a belief system within which you have a specific relationship with the Creative Principle.

I am going to propose that all the chapters of your life will be affected by the certain knowledge of it. I'd like for you to try it on like you were shopping for a new pair of boots. If it fits, then your life will be served by the adoption of it. If it doesn't, then you will have had, I hope, a pleasant walk. I trust you'll be welcomed to shop wherever else you will.

Ample evidence exists that we are participating in a quickening of the process that consciousness employs to descend into matter.

The upshot of this development in human evolution is that each one of us is also experiencing an acceleration in the process whereby our deepest feelings manifest the drama we call daily life. The dual nature of our world ensures that we will experience both the desirable and undesirable forms resulting from that quickening. Indeed, some of the monstrous creations that are coming forth from this process are the distressing subjects of the daily news.

If the root causes of our more monstrous creations are to be neutralized, a deep and universal self-examination of our personal motives must be undertaken. One of the most vital inquiries we ought to be making is the extent to which we are participating in the enslavement of others, no matter how subtle the process might be. This is a vital inquiry because the intent to enslave others becomes a self-fulfilling prophecy that has severe consequences for its authors in the age of quickening. I define the enslavement of another as requiring any person to use his or her precious time on the planet for activities that don't represent the highest and best use of his or her own free will to evolve, unfettered by the desires of other humans.

The fact that freedom is our intended state of being is amply demonstrated by all of nature. Yet we have conceived one clever way after another to ensnare and enslave each other to paradigms that no longer serve the evolution of humanity as a whole. While there are noble battles to be fought in the outer world, the most significant battle to be fought is not against the enemies you have been told of. The most significant battle to be fought is an internal one. Collectively, humanity is faced with the necessity to decide between two options for the future. The evidence is piling up that maintaining the status quo of bondage, whether voluntary or imposed, will lead to an involuntary and drastic reduction in the number of humans occupying the planet, an option most of us would prefer to avoid. The second option is to adopt an evolutionary paradigm shift and make a quantum leap into a new way of being.

Included in that quantum leap is the telling of a new version of the story of creation. We have each manifested in physical form in order to express our free will to create. We are each here to use our free will and to experience the results of the creative process that flows from the use of it. As we study the use of will, we remember, rather than learn, the very principles of creation that have always lain hidden just below the surface of our experiences. As we pursue mastery over the principles of the physical universe we will discover that the process continues in the non-physical realms of creation until we have finally merged back into the omnipresent consciousness of the Creative Principle itself.

The real truth is that each one of us, by design, is already master of our own universe. No one on the planet has manifested a biology that is identical to yours. No one on the planet has had the exact succession of thoughts and feelings that you have had. No one on the planet reacts to the stimulus that life provides quite the same way you do. You might not yet be able to accept the concept that you are responsible for creating the experiences that you call your life, but I'm sure, with a little coaching, you will be able to accept that each one of us has decided how to react to those experiences.

You'll learn how that reaction precipitates what happens next.

We have been designed as creators-in-training and have arrived here on Earth to discover that a self-taught on-the-job training course is in progress. The curriculum of the course is to understand all of creation. The methodology is to study the consequences of the use of free will and to use the resulting information to redesign the belief systems we hold until they come into alignment with the intent of creation. In later chapters I am going to provide the psychological tools and techniques you'll need to create an evolutionary paradigm shift. There are massive changes coming. The paradigm shift we need to create will help you to go forward fearlessly and experience those changes in a state of joy and gratitude. In order to fully describe that paradigm shift we'll need to build a common vocabulary around the main points.

So, on to the course work of exploring how the universe works in the context of our relationship with the Creative Principle.

Part I

A Story About Creation

Chapter 1

What's a Cosmology and Why Should We Create a New One?

Cosmology is defined as a "branch of philosophy dealing with the genesis, processes, and structure of the universe." A cosmology should answer the questions: "Who are we?" "What is our role in creation?" and "Why do things happen the way they do?" Ultimately a cosmology is a set of beliefs about the causes and consequences of our human experience here on planet Earth. It's a story we've been telling ourselves about how it really is. I want to tell a slightly different story about creation than the one commonly referred to as "consensus reality." It's a story that I hope will shed some light on why the world we have mutually created has turned out as it has. If the concept that individual and collective beliefs are the single most powerful force in creating our reality is new to you, then this book was written with you in mind.

The dawn of the global financial crisis in 2008 revealed serious structural flaws in nearly every method of political and financial organization on the planet. It also drew back the veil on a class of people that have been in charge of those methods. Millions if not billions of people have been watching their livelihoods being reduced to poverty levels, if not slipping away entirely. No one is pointing to any particular country and saying that they are the ones who have figured out a sustainable way to organize themselves. We've been through and rejected the tyranny of monarchy, dictatorship, and fascism. We've seen how the prohibition of self-interest in communism and socialism leads to apathy and disaster. So far the free enterprise system has been the best plan for the proliferation of goods. Responsible free enterprise provides opportunities for all. Irresponsible free enterprise provides an orgy of consumption for a few at the same time that it destroys the possibility of greater good for the many. The old cosmologies are crumbling. Where on Earth do we go from here?

We are beginning to see that humanity needs more than an abundance of material goods to be happy. What we crave is a sense of truth and meaning in our daily lives, a sense that we have the power to affect our destiny rather than being swept away by current events. Many think what is missing is a leadership that we can believe in. While better leadership would be nice, what we really need is a reason to believe in ourselves once again.

We need to reacquire the sense that what we choose to do each day, both physically and mentally, will have an effect on our individual destiny. In order to accomplish this, we need to rebuild our belief systems from the ground up. We need to create a new cosmology for the 21st century. None of the concepts we will use to construct a cosmology for the 21st century are new. In fact, some of them are so ancient they have been lost for millennia, which, in a sense, makes their rediscovery seem new. What might also be new for you is to see these concepts combined in this particular way.

It isn't really practical for most of us to apply the sum total of all the learning and wisdom we have each acquired in every situation that arises in our lives. If we did, there would be a lot of gaps in our experience while we wait for our biological "hard drives" to spool up and reacquire the original information we used to decide how we felt about a particular issue. Rather, we have grouped the decisions we have made about "how it is" into sets of beliefs about the world we live in. A belief is a mental act of placing trust in a person or concept, a mental acceptance of the truth or actuality of something.

Humanity is on board a spaceship with what appears to be a dwindling resource package. Yet, the universe is not running down. In an atomic sense, nothing really ever gets consumed. We just cause the atomic structure of matter to degrade from useful molecules into its basic elements without really understanding how to return those byproducts back into a useful form.

Despite our space-age technologies, we still don't realize that we are putting the survival of all humanity at risk by continuing to oppose each other, rather than learning to cooperate at a much higher level than we have demonstrated so far. One historic example of the clash of belief systems is the Cold War (1950–1985). The United States and the former Soviet Union spent billions upon billions of dollars preparing themselves to compete over the differences between their two political belief systems. Once these two nations were armed to the nuclear teeth, they spent billions more to destroy most of the capabilities they had just put in place.

The Soviets maintained that their belief system was based upon an appreciation for the dignity of the worker and the elimination of any sort of class system. But, in order to make sure there would be no class system and everyone would be equal, they put an elite class of privileged overseers in place to centrally manage the whole affair from the top down. Duh! Centralized control is always accompanied by institutionalized corruption. The result was generations of people who only understood how to take the direction of others and who had no understanding of the power of creativity. The whole construct lacked vitality at the grassroots level because the Soviet belief system failed to include a universal truth:

The personal expression of free will is a basic human instinct that affects the core motivations of all human beings.

With the collapse of the Soviet Union in 1991, the United States appeared to have won the controversy. The '90s were some of the most prosperous years in American history. We thought the future was bright and our brand of free-market capitalism was the holy grail of political ideology. After all, on the surface we had declared that the American way was based upon the free will of the individual.

The somewhat less shiny truth lurking underneath was that big business and our political leadership, under the direction of those who run our central banks, had long ago imposed upon the future expression of our free will by systematically draining the wealth of the nation to finance the building of their personal fortunes. They have created a system wherein they can take the profits, and the losses are absorbed by the taxpayers. They have secured the resulting national debt with the perceived value of American real estate and the full faith and credit of the United States Government. Who is relied upon to pay that bill on behalf of the United States? Why, we the people of course! Then our leaders engaged in every clever way they knew of to cook the books and obscure the fact that this version of the American Dream was beginning to suffer structural failure.

The straw that began to cause the camel's back to sag came in 2001. Its source appeared to be the clash of religious belief systems. Organized religions like to select a historical personality, celebrate his or her supernatural power, and declare their iconic figure to be the exclusive representative of God on Earth. They build ironclad belief systems around their concepts and interpret their scriptures to back them up.

Their members adhere to closed-loop belief systems as a matter of faith, and are unlikely to violate that faith by entertaining change. A closed-loop belief system is one that always returns to a single original premise. Ultimately, you either accept that premise and you are "in," or you don't and you are "out." They believe that the cosmology made up of their belief systems is the "only truth."

You may ask: If there are hundreds of cosmologies all claiming to contain the one and only truth about human existence, isn't our task simply to find out who is right and who is wrong? That argument would hold up if truth were a noun, a belief system to be possessed. I submit to you that truth is not a noun but a verb. We do not so much acquire the truth as we learn to practice it as a way of being, moment by moment. That kind of truth has been hinted at in the great scriptures of every major faith on Earth, and is itself as ancient as the "big bang," the massive explosion that occurred during the birth of our universe.

On September 11, 2001, we became painfully aware of the consequences behind the belief system of religious exclusivity when radical Islamists made several simultaneous kamikaze attacks on buildings in the United States, among them the Pentagon in Washington, D.C., and the World Trade Center towers in New York City. Is radical Islam alone in believing there is only one true path to God and too bad for the rest of the infidels? No, not at all. The practice is widespread, even here in the United States. Do the practitioners of religious exclusivity have a deep appreciation of the overall cost to humanity of such divisiveness? No, I don't believe they do. A significant percentage of military conflicts around the world are still based upon the clash between religious belief systems, much as the 9/11 attacks were. Religious exclusivity also discourages the practice of philosophical creativity, which leads to higher spiritual perception. It binds people to

> *"The budget should be balanced, the treasury should be refilled. Public debt should be reduced, the arrogance of officialdom should be tempered and controlled, and the assistance to foreign lands should be curtailed lest Rome become bankrupt. People must again learn to work instead of living on public assistance."*
> —Cicero 55 B.C.E.

closed-loop belief systems that do not match the intent of creation.

The conflict between domestic political belief systems is another way in which we are squandering our resources. Billions upon billions of dollars are spent each year to finance the debate as to whether it's the left wing or the right wing of the political body that has an exclusive hold on the true solution. Those who are watching closely are beginning to understand that all the hoopla over the Democrats versus the Republicans is only a drama to keep the general populace occupied while the biggest redistribution of wealth in human history is taking place behind the scenes.

The bill for the squandering of our resources is coming due. Once again it will be paid for by all common people everywhere and not by the politicians and central bankers who will not be held to account. They will tell us they are not the ones who created the basis for the current culture of self-indulgence with other people's money. They will assert that the system was already in place when they came to Washington. This assertion is true, but it ignores the fact that they haven't stood up to the existing system and made the changes that should be made on behalf of the people. Our real heroes haven't appeared yet.

Nobody will arrive at your home to present you with an invoice. The confiscation is underway and is taking place by remote control. First your 401(k) plan is devalued. Then, your stocks and bonds tank. The equity in your home disappears. Pensions go up in smoke. Wages and benefits are cut while the Congress wrangles on about raising taxes. Finally your buying power is reduced by an undeclared devaluation of the currency accompanied by massive inflation.

Political conflict has become a national sport. Your active participation in the illusion is implied and demanded. You are encouraged to take sides and told by the pundits and politicians that your vote is critical to the struggle—and indeed your vote is critical to keep the struggle going. Whether you choose to participate in the struggle is another matter.

The media circus that accompanies the clash of political belief systems is designed to keep you bonded to the belief that participating in the current "bi-polar" political debate is necessary to your survival. It is designed to distract you from looking outside the box. After generations of this nonsense, the general populace is finally asking why no significant and positive change ever seems to occur.

Science can also be bogged down by its belief systems. Science approaches its explorations primarily through the investigation of physical phenomena, but for the most part doesn't enjoy the spectrum of wisdom available from spirit. Members of the scientific community diligently keep track of each other's work, often updating their own viewpoints based upon the latest hypothesis proffered by their peers. Many avoid the topic of spirit, preferring to deal only with what can be proven in the physical realm. However, there are those within the scientific community who do recognize spirit and, I believe, benefit greatly from it. Of his processes of investigation, Albert Einstein once said, "I want to know the thoughts of God, everything else is just details."

Science is really good at describing what can be seen and measured. But, everything knowable cannot be seen and measured, and science is still more than a few experiments short of a viable "theory of everything."

Why do we need to create a new cosmology for the 21st century? We need to replace the belief systems that separate one culture from another, one ideology from another, one religion from another, and one discipline from another with a new worldview that emphasizes the universal concepts that unite us as a species. We need to do this before we squander what remains of our precious resources in opposing and manipulating each other. We need to learn to spend what remains of our terrestrial inheritance in a sustainable way, supporting and cooperating with each other before it's too late.

Wouldn't this new cosmology be just another belief system? Yes, it would. However, it is a belief system that recognizes the essence of our common humanity:

In the final analysis, we are all here doing the same things for the same reasons.

The structure of such a worldview must include what we know about our history, our planet, the Solar System, the Galaxy, and the interstellar medium it all floats in. It must include our best understanding of the intent of creation and our role in it. We are so wedded to outmoded belief systems that we might need to unlearn some things in order to learn others. It will be necessary to go back to basic principles in order to build a solid foundation under this new worldview of our role here on Earth. As we work together to create a cosmology worthy of the 21st century, we will build an awareness of the role of humanity in partnership with the Creative Principle.

Chapter 2

The Creative Principle

The word "cosmology" is made from a combination of the word "cosmos," defined as "the universe thought of as a systematically arranged harmonious whole," and the suffix "ology," which means "a branch of learning." By definition, the process of creating a cosmology demands that we turn our attention to the largest, most all-encompassing principles we can conceive of that might be responsible for the universe being a systematically arranged, harmonious whole.

Considering the enormous energy emitted by a single star, much less by billions of star systems, it is difficult to conclude that there isn't some sort of intelligent force at work in the universe pushing creation into being. In trying to describe that force, I am reluctant to use the word "God," which has been so abused by the religions of the world as to engender a different meaning depending upon who is doing the talking and who is doing the listening.

I began to actively investigate these issues in the early '80s. I had grown up in a small Midwestern town and went on to an education in design, graduating from college in 1971. I spent a few years living rather primitively in the mountains of Colorado. Then I founded a construction company that morphed from single-family home remodeling, to custom home construction, to single-family home development over the next ten years.

Eventually, I used my remodeling expertise to create a line of starter homes that were designed to be remodeled. Each small home came with the ability to be easily expanded in the future. We sized the heating and cooling systems for the larger home. We stubbed in the plumbing and electrical that would be needed, and framed in and covered the openings for future doors and windows. At the time inflation was running around 12 percent. That meant that a young couple could acquire a two-bedroom, one-bath home and in a few short years have the equity necessary to refinance and plug in the pre-planned additions necessary to keep pace with a growing family.

The concept was well received by the press and the real estate industry. The homes were reasonably priced little jewels of modern building technique. My partners gave me high marks for executing the project. I was 30 years old when I took out the equivalent of five million dollars in construction loans in today's dollars at the prime interest rate plus two points floating. Then something unprecedented happened. Over the next few years the prime rate went from 10 percent to 22 percent. I was faced with annual interest payments of 24 percent on five million dollars!

Needless to say, my dream collapsed. I lost the company, my home, the boat, and the airplane. Everything went up in smoke, including my self-esteem. It was the first crushing defeat I'd ever suffered and sadly not the last one. I slipped into hopeless despair. My marriage dissolved and my children started showing signs of being estranged from me. Later on I would learn that you don't get what you think you want in life. You get what you demonstrate in your heart that you are. But that realization was far in the future for me.

In 1983 I moved into a position as a project manager for a medium-sized shopping center development company and within three years was promoted to Vice President of Development. Temporarily it was a good fit. From 1983 to 1989 I produced nine shopping centers from raw dirt to grand opening in California and Nevada.

As a child, I'd been trained by my father as a craftsman and was still very competent around the workshop. In order to decompress from the rigors of the development industry, I worked with wood, leather, and metal in my spare time. I was in my mid 30s and had participated in the creative process of designing and manifesting artifacts, both big and small, thousands of times. Apart from the mundane aspects of the development business, I knew there was something spiritual going on. I was learning how to bring spirit into matter.

The prospect of having to spend millions of dollars every month and do it correctly, under time pressure, can be daunting. I was to go through the development cycle nine times in six years with a lot of financial amperage behind me. That's not a lot of power for some people, but I was having paranormal experiences doing it. Sometime in 1985 my company acquired a 19-acre site in a nearby town, which was one of the closest locations we had to our office.

One morning I drove up to the site, which was still a cow pasture, just to get a feel for the property. This would be my fourth project out of nine in the six year sequence. We had already completed a preliminary site plan and I knew the approximate footprint of the buildings. We hadn't yet decided what the architectural theme would be for the project.

I happened to have a folding chair in the car and I sat under a tree in my business suit, cup of coffee in hand, just gazing at the ground and trying to visualize the final product. I was toying with a red brick theme and was trying to construct a mental visual of the project. Then right before my eyes, what I can only describe to you as a mirage of the buildings in their final form, appeared. It lasted a few seconds and then disappeared.

Now, I was a child of the '60s and as Wavy Gravy once said, "If you can remember the '60s, you weren't there." I'd been around the block with perception-altering substances, but I'd never seen anything quite that visual using only caffeine for fuel. It got my attention. There were certain things I had noticed about the creative process I was engaged in. I knew that I was going to think about the project. I knew I would be making sketches of what I thought about. I knew I was going to speak about the project. And I knew I would be developing feelings about everything I was doing. That's all I was going to do. In a sense I wasn't going to do anything solid to make this vision appear in three dimensions.

Sometimes during this period I would unwind in the workshop by building a piece of furniture, say a chair. Usually, I do a series of sketches to round out the concept for a craft project and then I'll generate a detailed, drawing to work from. It dawned on me that I was building the chair as a preform within my consciousness and that my consciousness circumscribed the image of the chair and was distinct from it. I coined the term, "preform" to describe a non-physical holographic representation of whatever I wanted to manifest. I realized that the image of the chair was fully contained within the boundaries of my consciousness. I reasoned that the same thing would be true about the shopping centers I was working on.

In what we think of as the real world, I would be having endless meetings and conceptualizing every detail of the project with my architects, engineers, and salespeople. Together we would all contribute to a sort of shimmering preform of the project and record the history of our mental processes in the plans. I was running the creative process on a micro scale and a macro scale at the same time and beginning to see universal principles emerging.

I decided to launch an informal discussion group with my colleagues on the nature of the creative process. Our goal was to use the work we were doing to deduce anything we could about how the universal processes of creation operate. We would be talking about the principles that appeared to be operating within the Creative Process regardless of whether the project was in the workshop, on the development site, or out in the cosmos.

These were architects, engineers, and designers, all with a great deal of experience making roads, bridges, and buildings appear where nothing had existed before. We started the conversation by asking: "What is creative power?" We'd all grown up during the beginning of the atomic age. We all knew who Einstein and Oppenheimer were. We had all lived through the Cold War and the Cuban Missile Crisis. We were well aware that every atom of creation contains enormous power. We wanted to know where that power came from. If that was the same power we were using to create, then we wanted to have a better understanding of it.

Some of us were very religious, some were not. It became apparent right away that we couldn't discuss the power that pushes all things into creation in a cosmic sense without bringing the subject of God into the conversation. Right away we had a problem. Everybody had a different definition of what God might be. Some thought that God was a male and that he had human desires. He was prejudicial and liked certain groups of people more than others. He had a plan for each one of us and was harshly judgmental if we didn't learn to follow that plan. He was also benevolent at the same time.

Other people saw God more as a general force of nature. Then there were a few people that thought the universe evolved out of the chance collision of amino acids. Finally some folks had no concept at all.

Everybody acknowledged that the word "God" had so many different connotations that the use of a synonym or two might erase any charge on the subject and help to facilitate an evenhanded discussion.

In that pursuit we adopted the term the Creative Principle to represent that aspect of what God might be that is responsible for the generation of the physical universe. (I say that "aspect" of God to leave open the possibility that there is more to God than being the author of the physical universe. For the moment we wanted to focus on the processes relating to the manifestation of material objects). At any rate, the challenge was to find a fundamental concept, as a starting point, that we could all agree on. We managed to do that, and I present that concept here.

The Creative Principle is described by that set of information which not only includes, but circumscribes everything in creation.

The above statement was generated by the conversations we were having about how each one of us built a preform of whatever we were working on in our minds before it assumed physical form. We agreed that the consciousness that conceived an artifact circumscribed that artifact and yet was distinct from it. We reasoned that in the same way our consciousness can hold a preform within its boundaries there must be a larger form of consciousness that holds all of creation as a preform within its boundaries. That is what we wanted to investigate. In naming that entity the Creative Principle, we intended to depersonalize it and study it more as a force of nature, leaving mythology aside for the moment. In working with that concept we inadvertently gave birth to a paradox, which can be explained thusly:

The Creative Principle includes all of the atoms of all of the forms in the physical universe, yet its essential nature—being beyond form—has no form of its own.

This statement was generated by the knowledge that the consciousness that creates a preform and its subsequent manifestation is distinct from the atoms of its creation.

Was there anything in our direct experience that we could study that embodied these two statements? As it turned out, yes there was. If we remove all the free atoms of interstellar gas and dust, and all of the atomic material that makes up all of the star systems in the universe, what's left is a medium that once included all of the atoms of all of the forms in the universe, but, in its essential nature, has no form of its own. What that is, we call *"outer space."*

There is nothing else in the library of human experience that fulfills the two statements highlighted above other than the Creative Principle. Are outer space and our definition of the Creative Principle the same thing? I really wanted to know. I hadn't expected the conversation to go in this direction. But I had set this whole thing in motion and decided to go in whatever direction the conversation would take me. I had no idea it was going to lead me where it finally did.

Chapter 3

Outer Space

Outer space is generally thought of as a vacuum or absence of material, which is out there, but not here on Earth. How do you study nothing? I decided I could only study the nothingness of space by studying what it contains. We differentiate the concept of outer space from our experience here on Earth, where our atmosphere is clearly not a vacuum since you can't go anywhere on our planet where you don't bump into some atoms—even if it's only the ground you're standing on and the air you're breathing.

When I think about what all of the atoms of all of the forms in the universe might look like, the images from the Hubble Space Telescope immediately come to mind. If you haven't seen these fantastic pictures take a moment and Google the phrase "Astronomy Picture of the Day." An archive that is maintained by NASA will come up with a new and more spectacular image of interstellar space each day. Go to the bottom of the page and click on the word "Archive."

In the archive you will find images of great nebulas of interstellar gas and dust, plus one photograph after another of giant galaxies consisting of billions and billions of solar systems. You will learn that within sight of our instruments there are trillions of these giant galaxies.

The Milky Way galaxy (our home) is what is called a spiral nebula. A spiral nebula is a slowly whirling cloud of star systems with a luminous region at its center where millions of stars are concentrated. Our galaxy is believed to be 100,000 light years in diameter. That means it would take a beam of light traveling at 186,000 miles per second (that's 670 million miles per hour) 100,000 Earth years to cross just our galaxy alone. Our galaxy by itself is so huge the mind can't even grasp the thought. Never mind a universe consisting of trillions of galaxies of like size with untold space in between, expanding omni-directionally at a rate approaching the speed of light.

Since outer space is such a huge bite to digest, let's reduce our consideration of the concept of space to the size of our solar system to see if we can get a handle on it. In order to understand the scope and proportions of our own solar system, let's go out into the backyard and build a scale model of it using common objects that we find around the house.

For those not familiar with the process of scaling, it simply means that we select an arbitrary unit of measurement and reduce all the elements of a system by the same ratio so they still bear accurate proportions relative to each other. This is the same process used by architects to draw house plans so that an accurate image of a house will fit on a piece of paper and can be dealt with easily.

Let's say that we reduce the Sun to the size of a basketball. That's 9.39 inches in diameter. If everything else were reduced proportionately then the first planet from the Sun, Mercury, would be the size of the ball on your ballpoint pen and would be orbiting our basketball at a radius of 32 feet. So far so good; we're still in our backyard.

The next planet in line is Venus, orbiting our basketball at a distance of 60 feet. Venus would be nearly twice the size of Mercury, roughly the diameter of a number two pencil lead. Remember pencils? If you live in the suburbs of an American city, we are probably approaching the limits of your backyard already, but in rural Montana, where I live, our backyards are commonly much larger, so let's stay with the concept.

The next planet in line is ours: the good Earth. Our beautiful home, slightly larger than Venus, would also be about the diameter of a number two pencil lead and would be orbiting our basketball at a radius of 84 feet. Next in line is Mars. While the first three planets from the Sun have increased in size, Mars is actually about half of Earth's diameter, which would make it scale to the diameter of a mechanical pencil lead, the kind of irritating little pencil lead that breaks more often than it wears out. Mars is orbiting our basketball at a radius of 128 feet, so we are still inside our Montana-sized backyard.

Now we jump a little farther out to Jupiter, which is the largest planet in our system. Jupiter will orbit our basketball at a distance of 437 feet and is the size of a 25-cent jawbreaker candy—just less than an inch in diameter. Saturn takes us outside even a five-acre Montana backyard. At 802 feet from our basketball, Saturn, at roughly three-quarters of an inch in diameter, is like a medium-sized marble. No doubt I've just dated myself. You younger folks can Google the word "marbles" to find out what kind of toys your grandparents played with in their childhood.

Pushing on further into outer space, we come to Uranus and Neptune. In our model, both planets would be about the size of a sweet pea, the kind that Mom serves with dinner, orbiting our basketball at 1,613 feet and 2,529 feet respectively. We just departed even a Montana backyard and are now on a pretty good-sized ranch.

Finally, we come to poor Pluto. Pluto has been abused by the scientific community and declared to be "not a planet" any longer. Nevertheless, being reluctant to trash one of the foundation stones of the astrological movement, I include Pluto here to demonstrate where the outer limits of our model solar system might be. In our scale model, Pluto is about the diameter of a human hair and orbits our basketball at a whopping 3,324 feet. Impressive, huh?

However, so far we've only been describing how far away our planets are from the central basketball. In order to appreciate the true scale of our model we need to double all these radii to arrive at the diameter of each orbit. I need to specify here that all of the orbits of all of the planets around our model sun are actually elliptical and that the radii given are mean, or average, distances.

That having been said, the mean outside diameter of our model solar system as described by the orbit of Pluto (the diameter of a human hair) around our basketball is one and one-quarter miles. To put this in Montana terms, in order to draw an accurate model of our solar system on paper using the scale of the planets I've just described, you would need a piece of paper 800 acres in size. In the middle of that drawing you would place a circle 9.39 inches in diameter (our basketball) and then scatter a small handful of orbs into precision orbits at the distances specified above where they will each make their way around the basketball for millions and maybe billions of years.

There is enormous space between all these little dots. We can easily conclude that there is much more space in the universe than there is matter. You can clearly see that you could draw the basketball and each one of the scale planets on a piece of paper that is one square foot in size and still have room for asteroids and free atoms. So if we do the math in two dimensions it looks like this: 800 acres x 43,560 square feet in an acre = 34,848,000 square feet of space to one square foot of matter. When converted to a percentage this number tells us that our solar system is 99.99972 percent space and less than three one-millionths matter.

Now let's zoom back out to the full-sized solar system in which our sun is 870,000 miles in diameter and the average radius of Pluto's orbit is 3.64 billion miles. Remembering that our solar system represents only the tiniest dot within the confines of our galaxy, try to compare in your mind the ratio of the size of our solar system alone to the size of the average atom that makes it up. Unless you are an astrophysicist, and used to dealing in numbers with a whole lot of zeros after them, you are probably stumped.

The story doesn't stop there. In order to round out our concept of the nature of space we need to take a look inside the subatomic realm, too.

Chapter 4

Inner Space

Almost 2,500 years ago, the ancient Greek philosopher Democritus envisioned a world made of tiny particles. Our word for the atom came from the Greek word "atomos," which means "that which cannot be split" or "that which is un-cutable." Democritus theorized that if we were to cut a material in half, and then cut it in half again and again, eventually we would arrive at a tiny unit of matter that could no longer be cut, a unit of matter that was the essence of the original material with which we started.

Being ignorant of the processes of basic chemistry, our ancient Greek theorists proposed that each microparticle of a particular material would have the exact properties of the item they made up. Therefore pickle atoms would be green and lumpy and would taste like pickles, orange atoms would be segmented and squishy and smell like oranges, and so on. Democritus and his pals were seeking knowledge of the basic building blocks of the universe. If you favor the linear version of human history, they were some of the first theoretical physicists. In their quest for knowledge, they ran into the proverbial glass ceiling in that they had no scientific tools with which to pursue their hypotheses. Their work lay essentially dormant until the emergence of the Renaissance. When later on we take a look at the cyclic version of human history in Part 3, we'll see right away why that may have happened.

The Renaissance brought forth all the great thinkers featured in your middle-school textbooks: Leonardo da Vinci, Nicolaus Copernicus, Galileo Galilei, Johannes Kepler, and Isaac Newton, to name a few. These men, who were scientists and mathematicians all, had one other thing in common, they were all astronomers. They looked to the heavens believing that there they might discover the hidden clues that would explain why our world is the way it is. With the invention of the telescope they could peer successfully into the heavens, but when it came to the miniscule world of atomic structure they were still limited by the same problem that hampered the ancient Greeks: They had no tools with which to see into the sub-microscopic world. They were confined to experimental means in making conclusions about the nature of matter.

By the turn of the 20th century, the race to understand the atom was well under way. Still lacking the tools to actually see an atom, our scientists developed the language of mathematical physics with which to explore the unknown microcosm. In 1909, chemist and physicist Ernest Rutherford was able to use this language to establish the internal structure of an atom as being a dense nucleus of positive charge surrounded by layers of orbiting electrons of lower mass and negative charge.[1] That development ultimately led to the following realizations (which I am expressing allegorically).

First, the electrons orbiting our atomic nucleus have an electrical charge; and that charge generates a magnetic field. When two pieces of matter, such as a hammer and an anvil, collide, there is no actual physical contact between the steel of the hammer and the iron of the anvil. The force of impact is really the mutual repulsion of the magnetic fields generated by the electrons orbiting around the nuclei of their respective atoms.

You can see this force in action by trying to push the north poles of two permanent magnets together and observing the force of magnetic repulsion. You will be quick to note that you can't see or feel the magnetic fields of each magnet since they're not actually physical. So, we can conclude that what seems to be a hard shell around our atoms is not actually solid. Therefore, the impression we have that matter is solid is an illusion.

Second, as we investigate the nature of the inside of the atom we discover another astounding feature. Relative to the size and mass of the electrons, protons, and neutrons involved, we see the same kind of vast spaces in between the particles that we found in our visualization of the solar system in Chapter 3. The inside of an atom is composed mostly of space. As we probed deeper inside the atomic nucleus we found yet another layer of incredible space occupied by yet another magnitude of smaller particles. Still more evidence that matter is not really solid.

An understanding of the foregoing allows us to say that there is nothing separating inner space from outer space except the presence of minute electromagnetic fields. In fact, we can claim that inner space and outer space are one and the same, a continuous medium in which several different layers of oscillating magnetic fields are floating and, appearances aside, the physical universe appears to be an illusion.

Third, as we delve into the mysteries of the atom we find an additional phenomenon that is one of the keys to understanding the Creative Principle. In performing the experiments required to analyze the behavior of the most minute subatomic particles our scientists discovered that these minute building blocks of creation exhibited a dual nature: They alternately took form as waves and then as particles.[2] From this information some of our physicists have postulated the existence of a universal wave structure at a wavelength of about one-millionth of one-billionth of one-billionth of one-billionth of a centimeter. That's ten to the minus power of 33 zeros, one of those astronomical numbers we encountered while trying to visualize the contrast in size between an atom and our solar system.

What's the difference between a wave and a particle? Waves are not physical form, particles are. Let's go out to the stock pond in our Montana-sized backyard early in the morning while the surface of the water is completely calm. We know that the water is made up of molecules or particles. Pick up a pebble and throw it high into the air. When it strikes the surface of the water you will see that several small circular waves emanate out from the point of impact of the pebble in all directions. While you can reach out and scoop up a handful of water, you cannot scoop up the wave. The wave is not the water. The wave is an energy event.

You can look at the particles that make up a body of water and see where they are now. You can come back in five minutes and look at the particles and still see where they are. But if you look at the wave and see where it is, and then come back in five minutes, the wave is gone. The miniscule wave structures that precede the for-

The Creative Principle

mation of matter can't be predicted to show up in any one location because they are moving. They exist as waves of pure potentiality. Our scientists also found that waves carry and transmit information about the energy event that caused them to appear. The magnitude of the wave in our Montana stock pond could tell us the mass of the pebble involved and how high it was thrown. If all of this wasn't curious enough, while researchers were executing the experiments they devised to detect these smallest subatomic units, they found the units being studied tended to take form as particles when a human was monitoring the experiment, and as waves when no human consciousness was present.

This feature of atomic structure has been named the "Observer Effect." It has been said that the application of consciousness collapses the universal wave function and causes it to condense into a particle of matter. Science tells us that energy can neither be created nor destroyed. It can just be caused to change form. Through his famous equation $E=MC^2$ Einstein confirmed that energy can be transformed into matter and that matter can be transformed back into energy.

So here at the most microscopic level of creation we have a background texture of very miniscule waves of pure potentiality that have been manifested by the Creative Principle to receive the application of consciousness with which to transform them from energy into matter. As our story unfolds, we will see that our physical universe has been designed to respond to the qualities of individual and collective consciousness at all levels of evolution.

Every belief system eventually asks you to take a leap of faith.

The leap of faith I'm asking you to take is that the interaction between consciousness and the Creative Principle actually precipitates the seamless video that we call reality. This is a temporary leap of faith that only needs to be held long enough to verify it by direct experience. After that verification takes place, this leap of faith will transform from a belief system into a knowing. In later chapters we'll explore how this occurs.

Creation has been intelligently designed for us to explore the relationship between the quality of our consciousness and the resulting manifestation that we call life.

The Field.

Chapter 5

The Field

Let's do a little review. In our story about creation, the Creative Principle is described by that set of information that not only includes, but circumscribes everything in creation. In our story, the Creative Principle includes within itself all of the atoms of all of the forms in the physical universe, but its essential nature, being beyond form, has no form of its own.

There is only one thing in our experience of the universe that fits this criterion: space. Space actually permeates all of physical creation and is one continuous medium from the limits of the cosmos through the tiniest subatomic levels. The matter floating in space is essentially not solid. We found the smallest subatomic level of space populated by undulating waves of pure energetic potentiality. These waves demonstrate the ability to transform energy into matter as a result of the application of consciousness.

You may have heard this concept variously called the Divine Matrix, superstring theory, or the field of probability waves. I like the name used frequently by Wayne Dyer and Deepak Chopra: the Field of All Possibilities. Let's call it the *"Field"* to keep it short. The Field is actually the matrix upon which the physical universe is built. It is the fabric that underlies all of space, both inner and outer. It contains all of the atoms of all of the forms ever created. It contains all of the information ever generated. It is intelligent, interactive, and omnipresent, lying just under the surface of all creation. The ancients referred to it as *"ether."*

What could be the purpose for the existence of the Field? The Field was created so that the one could divide itself into the many. Why would the one want to divide itself into the many? That question takes us directly into spiritual mythology, and I ask you to remain aware that I am telling a story about creation.

The New Genesis

In the beginning the Creative Principle was all there was. The Creative Principle was self-aware, but there was no way for the Creative

Principle to experience itself from a viewpoint other than its own since it was all that existed. The Creative Principle conceived of a plan to divide itself into an infinite number of other units of self-aware consciousness so it could look back at itself from other viewpoints, and thus know itself and expand its consciousness.

Many religions tell us: In the beginning was the word. In some Eastern wisdom traditions the "word" is referred to as the sound *"aum."* In those traditions the word is the foundation of creation and all things that were made were made from it. Our concept of the Field is a more modern way to describe the biblical word. At the foundation of creation is a wave structure that has a wavelength so fine and so small that no other sound or wave structure within creation can ever be so tiny or so primal within this octave of creation. (The possibility that there are multiple octaves of creation within the physical universe is a concept that is beyond the scope of this book.)

This wave structure is vibrating at the highest possible frequency. As consciousness reacts with this wave structure, its frequency is slowed and the wave structure collapses converting pure energy into matter. Even though this wave structure is the smallest unit of creation, it also permeates everything in creation and is as vast as creation itself.

The Field was manifested by the Creative Principle to facilitate the emergence of the many, and to facilitate the expansion of the universe itself. Science tells us that the physical universe is expanding omni-directionally at a rate of speed that pushes the boundaries of human comprehension. The Milky Way Galaxy is the home of all the forms of consciousness we are aware of. By extension the galactic structures we see in the heavens are the home of all the forms of consciousness we are not yet aware of. The universe of galactic structures is expanding; therefore the intent of creation is the expansion of consciousness.

In our story, the Creative Principle is self-aware, but it has no form. Another way to say the Creative Principle is self-aware and formless is to say that it is composed *only of knowing*. Therefore, the first step in the process of the one dividing itself into the many was to create other individual units of self-aware knowing. To describe such an event is to push the limits of the written word. It is sufficient to say that when the Creative Principle conceives of something it is made manifest.

The very act of conceiving of the presence of other self-aware units of knowing caused the Field to take form so the many would have a place to come into being, indeed so that the very concept of place could come into being. This process of self-division can also be called the *"creation of souls"* (a soul being defined as any individual unit of self-aware consciousness).

To create an infinite number of souls of pure knowing is to create a universe that does not yet consist of physical form. Creating an infinite number of viewpoints that are all the same would not facilitate the expansion of either a physical or non-physical universe of infinite variety, which is what we see around us. The very nature of the division of the one into the many gave birth to free will so that all souls would have the right and ability to know the Creative Principle from whatever viewpoint they chose. The free will to create is the single characteristic that identifies all souls as being made in the image of the Creative Principle.

This universe of knowing is *the universe of first cause;* meaning all things that have appeared emanated from there. Free will is the reason that creation has a wondrous variety and beauty in it. The universe of knowing (being divided into individual embodiments of knowing) has come to be called the *"causal universe"* or the *"causal plane."* These individual bodies of knowing, these souls, are the off-spring of the Creative Principle. For the first time the Creative Princi-

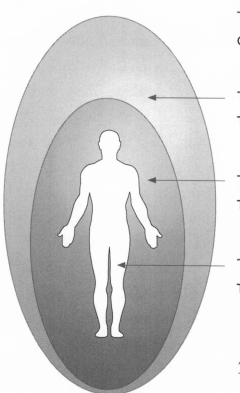

The Creative Principle
Circumscribes everything in creation

The Causal Body
The realm of knowing

The Astral Body
The realm of thought

The Physical Body
The realm of feeling

The multidimensional human.

ple was able to know something of itself from "there" instead of from "here." By now the Creative Principle was aware that it could conceive and that it was having ideas. The Creative Principle wished for its offspring to be able to have ideas as well to facilitate thinking about itself from different points of view.

The act of thinking is distinct from the act of knowing. To borrow a concept from Democritus, knowing is that which is indivisible. That which you have declared you know is not as subject to revision as that which you are simply thinking. As the Creative Principle conceived of offspring that could think, the universe of ideas came forth. We have come to call the universe of ideas the *"astral universe"* or the *"astral plane."*

In myriad ways, the individual conscious embodiments of knowledge existing in the universe of first cause can descend into another universe of slightly denser embodiments of thinking. There, in the realm of pure light, astral souls can manifest any idea they wish. As in a dream, the astral universe is highly malleable and so-called "reality" can be easily manipulated by the process of thought.

The Creative Principle wanted the process to continue, and so it did. Knowing and thinking are distinct from having the direct, personal experience of feeling. The Creative Principle wished to extend its senses into a physical world to feel what it was like. But the Creative Principle had no form with which to feel. As the Creative Principle conceived of the idea of feeling, the *physical universe* came into being. The birth of the physical universe, also known as the *"physical plane,"* is consistent with what our scientists refer to as the *"big bang."*

The physical universe, where we live in our physical bodies, is the realm of feeling and of emotion. It is here that the Creative Principle can feel its creation. How can the Creative Principle extend its nerve endings into the physical world to experience itself, if it has no form? Quite simply, we are the nerve endings of the Creative Principle. We are the sense organs with which the Creative Principle feels its own creation through the Field. All biological forms are the nerve endings of the Creative Principle. The reason biology exists is so that the Creative Principle can sense its creation. This point will become relevant to our later discussions because it is through the language of feeling that we are able to hold a conversation with the Creative Principle.

Having descended through two subtle universes to reside in the physical world, each of the children of creation has a physical body within an astral body within a causal body.[1] The causal body is very subtle and has the faculty of knowing. The astral body is slightly denser and has the faculty of mind so it can think. The physical body is still denser yet, and has organs of sense so it can feel.

We, the children of creation, are made of the very same stuff as the process of creation and are thus made in the image of the Creative Principle. Each of us has the same power to create through free will as the Creative Principle itself does. This feature of creation has not been suggested, and it is not something you win the ability to do through good behavior, or by following the rules. It is mandated by the very structure of creation itself. You do not have the option to participate. You cannot stop your process of thinking and feeling and still remain in physical form. You must participate by design.

The Field contains all atomic structure. The Field flows through everything in creation, and all of creation is floating in it. The Field connects everything with everything else. The Field is the nervous system that connects the Creative Principle to its sense organs (us), and the way that it senses everything that goes on within the three spheres of existence.

It is common to hear it said that "we are all one." The idea of the Field provides us with a concept grounded both in physics and in myth that shows how we actually are connected. The truth is that we're much more connected by the Field than we are differentiated by the unique expression of our individual atomic forms, which, in their very essence, are not really solid, but are electromagnetic manifestations of free will floating within the cosmic body of the Creative Principle. Since we have all descended from the one consciousness that circumscribes everything in creation, the universal truths regarding our origin and our purpose are hidden somewhere within the confines of our individual consciousness. You may have noticed that the act of manifesting in physical form obscures the knowledge of your previous existence as an eternal unit of self aware consciousness. That is why I'm saying that the task before humanity is to *remember* (rather than to learn) who we really are.

Consciousness flows from the subtle to the gross. As consciousness descends through each of us it creates an electromagnetic cloud of intent that interacts with the Field to collapse the wave function, both attracting and manifesting the particles of matter that make up the world around us. How this descent of consciousness from pure knowing into physical reality manifests our world and our lives is the essence of our relationship with the Creative Principle.

Chapter 6

The Law of Attraction

Each person walks upon the Earth as a subset of that set of information that makes up the Creative Principle. It could be said that you inhabit your own individual universe comprised of that small subset of experience, identified as you, which is contained within the larger set of experience identified as the Creative Principle. You are having a unique experience.

No one else on the planet is enjoying the same series of thoughts and feelings in the same order as you are. Just as the overall universe is growing, so your universe is also growing.

Each act of will you undertake adds to the growing body of information that is included within your universe. Each thought you have, each emotion you experience, each action you take, and each realization you have are all acts of will. Every act of will is also an act of creation. Every act of creation you undertake expands your universe. The fact that your heart is beating is an act of will to exist in the physical world. As you lie in bed at night, your body is actively creating new cells with which to maintain your physical presence in the world. You can't get out of bed and brush your teeth without performing a series of acts of will. Each act of will contributes to a statement of what you believe to be true that resides in the deepest recesses of your being, and, as such, sends waves of intelligent instruction coursing through the Field. You don't necessarily get what you think you want in life. You get what you have demonstrated in your heart that you are.

There are two ways to focus repetitive acts of will. In one paradigm, reality is thought to be a series of events that are caused by going out into the world and pushing atoms around to force the forms that we desire to appear in our lives. The focus is on what exists on the outside of a human. Many people have chosen to worship the acquisition of material objects and material-based security as their main purpose in life. Popular American culture celebrates the accomplishments of the rich and famous as examples of what we should all hope to become. At the same time popular American culture depicts such people as being hopelessly mired in drama and unhappiness.

In the other paradigm, reality is thought of as a series of events that are drawn into experience by the attractive power of the consciousness within an individual. The focus is on what exists on the inside of a human. Central to this theme is the concept of the spiritual master. In India, a spiritual master is considered to be one who has achieved mastery over consciousness and is immune to the restrictions of the physical world.[1] Indian culture abounds with stories of those exalted souls who are capable of causing atomic structure to assume any form they wish. They are capable of performing what we think of as "miracles." Western Christianity is based upon the story of such an individual. Having acquired such knowledge and power they no longer pursue the accumulation of material objects preferring to focus their attention on the qualities of consciousness itself.

These paradigms are, of course, two extremes; and most of humanity holds beliefs that lie somewhere in between. What do two people embracing these very different paradigms have in common? Each person engages in a lifestyle of highly-focused acts of will. Both have learned to consciously operate the Field to manifest what they want. Knowingly or not, they are participating in the same process. Knowingly or not, all of us are participating in the same process.

Creation in your life flows from the subtle to the gross because the Creative Principle manifests the material universe from pure consciousness. The process of creation begins in the most subtle realm of first cause, the realm of pure knowing, which has been totally and completely authored by you. It is that set of core beliefs that are reflected in the quality of your ideas, which appear seemingly out of nowhere and that make up your individual cosmology. Most people think their ideas just pop into their heads. If you are aware, you'll have noticed that your ideas actually result from a series of non-verbal acts of will.

Just before the appearance of an idea, a choice is made as to both the content of the idea and the feeling that goes along with it. Once you embrace the concept of living life from the inside out, you'll understand that the emotional choices you make, as an idea is formed, constitute the most subtle level of creation and also the most powerful. It's also the most difficult layer of your consciousness to access. Everything that happens to you begins with emotional choices that occur on a level that is so subtle you may not be aware that the process of making those choices is something you are actually doing. The goal is to become more aware of what you are doing on a microscopic level.

The Creative Principle experiences each one of us as a creator-in-training—each manifesting our own unique, but partially overlapping universes. The Creative Principle honors every input we make into the Field as if it were the word of a god. Every one of the thousands of choices we make, just before the expression of an idea, amasses a cumulative electromagnetic cloud of emotional intent in the causal plane. This electromagnetic cloud of intent forms your body of knowing and a declaration of what you believe to be true. Your body of knowing acts as a powerful magnet that has strong attractive powers. That power of attraction draws the willing players in life's drama to each one of us to fulfill that which we have declared to be true. Furthermore, each one of us operates this cycle thousands of times per day, constantly supporting and adding to the body of knowing that manifests our reality. It's as if each of our thoughts and emotions is represented by a single grain of sand flowing through an hourglass. At first look, the sand appears to be flowing at an insignificant trickle. But if we come back tomorrow, the whole room is full of sand.

The Creative Principle accepts every input we make as if it were truth because it comes from that part of us that knows. That is not to say that our body of knowing necessarily corresponds with any sort of universal truth. Everyone's body of knowing is completely subjective. Every thought that is repeated consistently becomes a belief. Every emotion that is repeated consistently becomes a belief. Your body of knowing is made up of your innermost core beliefs. It consists of all the choices and declarations you have made in your life. It consists of what you have told yourself that you know to be true. The purpose of the Field is to honor those choices and declarations by manifesting them.

Therefore the Creative Principle, through the Field, manifests your word as a statement of what is true in your universe. This means you have the free will to create your universe in any way you want, and to experience whatever monsters or whatever joy you wish to conjure up. Each one of us can change the reality we live in by changing what we think, what we say, what we do, and how we choose to feel about it all. Why do we not realize this? Because there is a time delay between when those acts occur and the manifestations that result from it. That time delay has been included in intelligent design lest the universe be destroyed by a single thought.

If my horse does something I don't like, I can't come back tomorrow and discipline her to make the point. She won't get the connection that the discipline she's receiving in the here-and-now is the result of something she did yesterday. Likewise, we humans don't seem to get the connection between what we've generated internally in the past and the life that we experience in the here-and-now.

Part of the difficulty in creating serious change in our lives comes from the fact that our body of knowing has built up over time. If a momentary inspiration about the changes we'd like to make were a bottle of dye, we seem to think that tossing that little bottle into the ocean of knowing we have created will somehow change the color of the water. You see the difficulty. In order to effect serious change in our lives we must be absolutely diligent to see that every input we make into the Field is consistent with the reality we want to manifest.

One member of my family smoked cigarettes for 50 years. Finally he recognized that the habit had the potential to kill him. One day he declared out loud, "I don't smoke, I have never smoked, and I have no intention of starting now." With no knowledge of the Field or any of the principles we have been discussing, he was absolutely consistent with his declaration. One day he just stopped smoking. He didn't have the struggles or the withdrawal that we associate with quitting. There was no drama. One day he smoked, the next he did not, and he has never gone back. I'm not recommending that denial of the past is the best way to proceed. However, this story is a prime example of making change through the power of free will.

Let's say you'd like to manifest prosperity in your life. Ever since you can remember, having enough money has been a struggle. You are sick and tired of worrying about the bills and where the next dollar is coming from. Your life coach tells you to spend one hour per day sitting in a chair visualizing what it would be like to live in abundance and be free of worry. You keep throwing that bottle of dye in the ocean and yet the color of the water doesn't change. Why not? Because you still spend the other 23 hours of each day filling your body of knowing with the emotion of worry about how you are going to meet your obligations.

Remember that through us the Creative Principle extended its senses into the physical plane to experience how it feels. The physical world is a world of feeling. That being its primary purpose, the

Creative Principle doesn't so much manifest from the subject matter of your thinking as it accepts the collective body of your emotion as your word and the truth about how life is in your subjective universe. It then plays your life back to you demonstrating exactly what you have declared that truth to be.

Plays your life back to you? In a manner of speaking, consciousness is the light source that projects intention onto the movie screen of the Field, causing atoms to coalesce and form an image of so-called "reality," which is essentially non-physical. This miracle of creation is what we term "daily life." It is you making phone calls, going to the store, driving your car, relating to your friends and family, and recording how you feel about the whole experience.

The cooperative components that make up the seamless video of your life are attracted to you from far and near to fulfill your personal statement of what is. That is divine law. Your life situation and the current condition of the world have resulted from what each one of us has done with this miraculous mechanism. The real miracle is how we could make something so fantastic seem so ordinary.

Part II

The Dynamics of Being Human

Chapter 7

A Quantum Leap

There are more than 200 billion stars in the Milky Way Galaxy. So far we only have knowledge of one star system that supports life. Earth occupies just the right orbit around the Sun, resulting in just the right range of temperatures, for life to have flourished here in ever more complex forms for the last two billion years. It is estimated that there are 500 million planets in similar habitable zones throughout the galaxy. Strict materialists love to postulate that the process of evolution explains the development of all species with no need for a higher power to drive the whole affair. But there are "pesky gaps" in the fossil record that can't be explained by linear evolution. For example, there is no fossil evidence of a bridge between Neanderthal man and *Homo sapiens.*

It seems that the process of evolution isn't seamless. Most people investigating the fossil record in detail know that there isn't a smooth and gradual flow of intermediate forms between documented fossilized species. In fact, the fossil record is made up mostly of those pesky gaps, for which we have no explanation,

Immediately after naturalist Charles Darwin published his findings in *On the Origin of Species* (1859), the overall sentiment was that we just needed to search more and we would eventually find all the fossils we needed to fill in the gaps in the record and produce another nice constant we could rely on to explain our existence on Earth.

Well, scientists did search, and mostly what they found was more gaps. Today, the debate rages on with evolutionists on one side claiming that once we've completed excavating the entire planet we'll see that life has evolved from a random recombination of atomic elements through the process of natural selection. On the other side, we have Creationists claiming that the fossil record is no evidence of anything, and that God created man only about 5,000 years ago. As in political debates, the two sides are committed to mutually exclusive theories, and very few are looking outside the box.

Why do the two sides of this debate vehemently oppose each other? The answer is simple. Each side has built enormous institutions around its own point of view. The people who make their living traveling around the world excavating here and there for the holy grail of evolutionary evidence wish their activity to continue. The people who preach strict creationism from the pulpit wish their activity to continue. Both maintain that their respective positions represent the only truth. If the problem were solved by proving a third distinct possibility, two engaging ways of living might become obsolete. So they cling to their positions while the evidence builds up that the real truth is outside of both boxes.

Here's another possibility, a new, more inclusive definition of the development of our species.

What we think of as evolution is only the speed at which the Creative Principle conducts its experiments in the embodiment of consciousness.

If this is so, then how do we explain those pesky gaps in the fossil record? Does evolution have more than one speed? Does time have more than one speed? Could it be that every once in a while the acceleration of a particular strain of consciousness reaches a sort of critical mass and spontaneously makes a quantum leap into a new expression of itself in the physical plane? Could it be that this process is a regular occurrence, common to all species over time, which is written in the divinely inspired instructional manual of our DNA? For the purposes of our discussion a quantum leap is an energy event wherein an individual embodiment of consciousness experiences a distinct transformation both spiritually and physically. When Jesus healed the sick, he caused a quantum leap to occur within their physiology. His message wasn't that he was the only one who could accomplish such a feat. His message was that all of humanity can learn to do even greater things than these. One of the promises of the Creative Principle is the possibility that not only humanity, but all sentient beings, are capable of instantaneous transformation.

Let's back up a little and see if we can put some legs under this concept. If we were arrogant Westerners, we might naturally assume that the Field of All Possibilities was designed to react only with human consciousness. But if that were true, it wouldn't be the Field of All Possibilities would it? Since we are trying not to be arrogant Westerners, and we have an innate respect for all forms of life, we're willing to explore how all life enters into the mix.

The human body is a composite structure. It has roughly ten trillion cells in it. The human body is also home to billions of microorganisms. These minute biological forms are born, experience desire, defend themselves against invaders, live out their lives, and finally expire, just as we do. They are animate. They move around and do their "jobs." They don't enjoy the cognitive functions that we do, but we couldn't experience the cognitive functions we do enjoy without them because we'd be dead. They perform critical jobs necessary to the biology of a human body.

Desire indicates consciousness. If you've ever tried to make serious changes in your diet, you've likely experienced the collective desires of the hoards of microorganisms living within you that you have trained. The cravings that you feel are the result of all of them banding together and yelling, "More pizza!" or "More sugar!" Or yelling for whatever else you have trained them to be addicted to. Of course, they do this by sending chemical signals into your nervous system. But they might as well be hollering at you, as the result is the same. The extra weight so many of us are carrying around our middles demonstrates that we tend to pay attention to what masquerades as a craving, which is really a nonverbal message from another life form.

There may come a point at which our microorganisms get out of balance. When that happens, our temperature rises. The body may excrete various fluids and even quake for a period of time. If we can't heal ourselves easily and naturally, we may take steps to kill off the little buggers that are disrupting our lives until we regain our balance. Consciousness exists in fractal layers within creation. A fractal is a geometric shape that can be split into parts, each of which exhibits the same patterns as the whole regardless of its scale. As above, so below;

as within, so without. The principles that make up our world also exist in the same universal patterns regardless of scale. As the human body is a composite being, so is our beloved planet. As the human body is host to billions of microorganisms, so the planet is host to billions of units of consciousness smaller than itself. That not only includes us, but all of the other species on Earth.

If the Earth's micro-beings get out of balance, her temperature may rise, she might excrete various fluids, and even quake for a period of time. If she cannot heal herself easily and naturally, she may take steps to kill off the little buggers (us) until she regains her balance. This is the same process that our bodies go through with our microbial inhabitants. As above, so below. Just as we are willing to allow legions of microorganisms to inhabit our bodies, the Earth is also a willing host for the human free will experiment.

In 1979, environmentalist and futurologist James Lovelock published a book entitled *Gaia: A New Look at Life on Earth* (Oxford University Press). During the early 1960s, Lovelock worked for NASA developing sensitive instruments for the analysis of extra-terrestrial atmospheres and planetary surfaces on behalf of the space program. In so doing, he discovered that the Earth employs a conscious, self-regulating feedback loop to keep conditions on the planet safe for the proliferation of life. During his research, the name Gaia (after the Greek goddess of the Earth), was given to this function. One of the first things stated in the book is: "The quest for Gaia is an attempt to find the largest living creature on the Earth."

Lovelock's Gaia hypothesis proposes that living and non-living parts of the Earth form a complex interacting system that can be thought of as a single organism. The presence of a living organism, of course, implies some level of consciousness. Despite overwhelming evidence to the contrary, many would have us think that our planet is just a chemical/geological phenomenon that doesn't harbor any sort of awareness. It's just rocks and water. Lovelock makes a compelling case for planetary sentience.

Being completely self-centered, part of the problem we humans have in considering the concept of consciousness is that we fail to allow for the existence of a spectrum of consciousness. The animal kingdom shows us a different level of consciousness than ours. The microbial kingdom shows a different level of consciousness than ours. So does the subatomic realm. So indeed does the Field. Why would we assume that something as magnificent and complex as a planet wouldn't enjoy some sort of self-awareness? One of the reasons is simply this: If we limit ourselves to what can be proven within the confines of our test tubes, by only studying what exists in the physical world, the entire non-physical universe, by definition, does not exist. According to some spiritual masters, the non-physical universe actually dwarfs the physical universe in size.[1]

Consciousness is seeking to express itself through a range of manifested forms from the very simple to the highly self-aware. The Field is designed to interact with consciousness on all different levels. Every conscious entity at whatever level has been granted the divine right to self-determination. The Field awaits the bidding of all. As the Bible says, "Knock and the door shall be opened unto you. Ask and it shall be given."

The Creative Principle has only one commandment for you:

"You shall create."

Your life represents the essence of this commandment. Your life is an experiment in consciousness manifested through the god-given right of free will. All forms of consciousness at all levels are pursuing self-knowledge in the quest to expand awareness to the point where it approaches that set of information that not only includes, but circumscribes everything in creation. One of the basic beliefs of our new cosmology for the 21st century is that, by definition, the Creative Principle grants the free will to explore self-knowledge and transformation to all.

As a species, how are we doing so far? Let's work it backwards. The state that our world is in is the direct result of the operation of this magnificent and elegant system of manifestation by us—a system that is entirely in balance at all times regardless of what we use it to manifest. What is happening to each one of us is exactly that for which we have called. The future is not assured. It depends upon the actions of our mutual free will. The experiment is all about how we're going to adapt to our current predicaments. Have mass extinctions of self-aware species occurred in the past? Yes. Will they occur in the future? Maybe. If it does happen again, does that mean we will cease to exist? No. Energy cannot be created or destroyed; it can only change or be caused to change form. If we perish physically, we will retreat back into our astral bodies as we have many times before.

By virtue of the fact that there are unprecedented numbers of people on the planet, we find ourselves in uncharted territory. We are beginning to strain the natural systems we rely on for our physical existence, perhaps beyond the will or the ability of Gaia to adjust and rebalance. Cultural and economic systems are collapsing around the world. There is a component of acceleration in the affairs of humanity. Can you feel it? Time seems to be moving faster and faster. Current events seem to be leading us in a particular direction. Those that are looking beyond consensus reality can see that the forces carrying us toward the possibility of extinction are the same forces that could be carrying us toward the possibility of transformation from a species being swept along by its own ignorance to a species showing a new level of mastery over itself. That's what happens just before a quantum evolutionary leap occurs. That is why it occurs. Evolve or perish.

For the first time in recorded history, we actually possess the technology to destroy ourselves. If that isn't a manifestation of consciousness, what is? Shakespeare's question "To be or not to be?" is being called out by the collective actions of all humanity. Will we be able to step up to the plate and do the things that are necessary to end the madness of hate and greed, and continue the coursework of creation? Even though we are surrounded by messages telling us that the old paradigms are obsolete, will we try to put it all back the way it was? As we explore the concepts of the Creative Principle in the remainder of this book, we will discover that our cultures are collapsing because the paradigms that manifested them are unsustainable self-fulfilling prophecies. We are finding out quite literally that we can no longer afford to make war upon each other. Those who are thinking ahead are finding out quite literally that we can no longer afford to exploit each other for personal gain.

Every political and economic system on the planet has a party line. The party line is what they say they are doing. But the Field is recording and manifesting from what they are actually doing in their hearts and in their minds. That is why there is such a huge disparity between what's being said and what's actually happening. The players in the game of politics are about to file a massive bankruptcy on both moral and fiscal grounds. The old paradigms are no longer viable; it's time to change the game internally. Once we've done that it will begin to change externally.

Clearly, an acceleration in human affairs is taking place. In Part 3, we will look at evidence showing there may be an astrophysical component to the acceleration we are all feeling. The quickening is upon us, so it feels fair to ask: Are we about to manifest a true reflection of our human potential, the ability to live in a natural state where every sentient being enjoys the freedom to pursue higher awareness and self-knowledge? Is the human "train" on rails and headed directly for an abyss, or are we poised at a level of critical mass that will propel us to make a quantum leap developmentally, transforming ourselves into the next evolutionary stage of *Homo sapiens*? Once we have embraced the fact that our world, as it is, has resulted from our collective operation of the Creative Principle, then we will have an effective place from which to alter our cosmology and create positive change in the future.

Chapter 8

Who Is Judging Whom?

In his third law of motion, Isaac Newton expresses the idea: For every action there is an equal and opposite reaction. While Newton was concerned with the reaction of two physical bodies upon each other, his third law may also hint at the presence of a deeper spiritual principle, which can be expressed:

Every act of free will creates an emanation of consciousness that flows outward from its source and interacts with the Field, resulting in a manifestation that flows back toward its source in an equal measure.

When translated into behavioral terms, this principle doesn't appear to mean that if you harm another person you will be harmed in an identical manner. We've all seen examples of people who have done grievous harm to others and walked away from it. Religion attempts to deal with this apparent inequity through the concept of divine judgment. The conventional Christian view is that we should always keep the concept of divine judgment before us and behave properly for fear of what may happen to us if we do not. Those who believe in an afterlife, for instance, say that we might not experience the payback—rewards and punishments—in this life, but we will surely experience it in the hereafter.

It's important to note here that we accept this because we are en-couraged to be afraid of how we'll *feel* when divine judgment comes. The belief system that has been built around divine judgment is based upon the idea that whatever happens, it's not going to *feel* good and we should take great care to prevent that from happening. A large segment of the planetary population has accepted the fear of divine judgment as the basis for a fundamental sense of right and wrong. This group would loathe discarding that paradigm under any circumstance. One might ask how the Creative Principle can endow every soul with free will and then judge each one according to its use? By what criteria does that judgment take place? Some faiths answer these questions by specifying what the penalties for certain actions will be, others declare that the ways of God are mysterious and we can never hope to under-stand them.

I don't believe the Creative Principle is motivated to remain a mystery or hide the principles of creation from us. On the contrary, I think it is becoming ever clearer that the intent of creation is that we expand our consciousness in every way possible. The curriculum of human experience is all knowledge. If we can accept the premise that Newton's third law not only applies to physical objects but also hints at a deeper spiritual principle, we should be able to see that the sys-tem of creation is balanced because every act of free will produces an equal and opposite reaction in the Field. Could it be that in terms of emotional energy, you get back exactly what you put out? While the system itself is both balanced and elegant, the end result, the reality that we are manifesting, is not balanced; but that has to do with the input, not with the system itself. As computer programmers like to say: Garbage in, garbage out.

If you harm someone, you will not necessarily receive the same outward experience (although you could), but you will attract into your life a cascade of events that will produce an emotional content of *feeling* equal to that generated during the course of the original act of creation. How does this come about? The anger, the hate, the fear, the carelessness, or the disdain that fomented an original event is recorded in the Field. The Field interprets the information it receives from our thoughts, feelings, words, and actions, as a statement of "what is." Since the information was declared to be true by a being with the power of a god, the Field is obligated to respond to this conscious input by reflecting an event of equal emotional content back to its source. The reaction is opposite in that the original emanation traveled away from its source and the resulting manifestation is returning to its source in the opposite direction.

Here comes the *humbling crew*! The humbling crew is that group of people who have been drawn onto the stage of your life to help you act out your "screenplay." They have been drawn to your "drama" because their body of knowing has something in common with your body of knowing. The victim seeks the perpetrator. Their purpose is to show you what is actually in your script. Humility begins to arise naturally once you become aware that your every thought and feeling is being displayed on the silver screen of life.

Let's draw a distinction between two ways to look at life. In the first view, your self-image is one of a supplicant being judged by an outside force, according to a criterion that hasn't been clearly described. You don't know how you will be judged or what the punishment might entail. You live your life in the risk that you may have deluded yourself about how you are doing, though you won't really find out until the Day of Judgment. The ways of God are mysterious. This view precludes you from being a creator-in-training and relegates you to the role of sinner in dire need of forgiveness. In this first view, you only have one life to live. You become a "sinner" simply by being born in a physical body. If you have willfully disobeyed God's laws and do not repent, you will be punished for eternity—no do-overs. But in a compassionate universe, punishment is supposed to be seen as something metered out to help you change your behavior and improve yourself in the future. Eternal damnation with no do-overs has no value to anyone because it leaves no opportunity for future redemption. If the intent of creation is to expand consciousness then divine judgment cannot exist.

In the view I'm offering you here, your self-image is one of a creator-in-training. The Creative Principle will give you back exactly what you put out until you *see the connection* and begin to change your behavior, and consequently your declaration of what is real. At that point, you will still get what you put out, but you will probably like it more, and that is the crux of the matter. In both views the participants will tend to adhere to the principles of right living, but for different reasons. In the first example the participants are motivated by fear and a desire to please a higher authority. In the second, they are motivated by the joy of raising their consciousness to a more celestial level.

The next concept we want to add to our cosmology reads as follows:

The Creative Principle does not judge you, but has designed a system wherein you pass judgment on yourself with every act of willful creation in which you engage.

We all know that the physical body is temporal; it comes and goes. When the physical body has returned to dust, the consciousness that once inhabited it still remains within the astral and causal bodies that were associated with the soul that formed the physical body. Thus, after your physical body disintegrates, your consciousness is not extinguished and you continue to have an opportunity to evolve. The astral you might want to continue to experience the world through feelings and emotions, so you may descend into the physical plane again and again to fulfill your desire to experience yourself in the realm of feeling. Understanding the course work of creation can take more than one lifetime.

What a wonderful system! Everything that happens to us is a result of actions we have taken. What has already happened to us has already happened and can only be changed by its relationship to what happens next.

I could dwell on all the horrible things that have happened in my life in the expectation that the future will bring more of them and I would be correct in that assumption. Or I could choose to see that all of those experiences have added to the wisdom I have acquired which is helping me to reshape what does happen next in a fashion that is more to my liking. I would be correct in that assumption as well.

Our reaction in the here and now to what has already happened will be partly responsible for what happens next. Once we have integrated this knowledge, it's not hard to see that there can be no more blame assigned to others for the negative events that we are experiencing.

If you believe that everything that happens to you originates in acts of your own free will, how can you point a finger at the next person and accuse him or her of causing the experience you generated? If you believe that what happens you are doing to yourself, you will take much more care when deciding how to react to life. You see how we can be motivated to live a moral life without the need for divine judgment.

It isn't difficult to understand that blame and retribution is at the heart of every conflict that is occurring in the world. Like the feud between the Hatfields and the McCoys, some conflicts have been going on for so many generations that the current participants can't even remember why they are so angry at the people on the other side of a given issue. They do know that they are angry, and that they can't find the strength in their hearts to forgive and move forward.

If you keep a sharp eye out, you'll notice that the people who believe in divine judgment are often the first people to lay judgment upon you. They don't realize that the act of judging others will cause the essence of that which is being judged to appear in their own lives. We would all do well to remember that the process of creation causes us to become what we judge in others. This is a part of divine law, and is the true meaning of the proverb: "As you sow, so shall you reap." We've established that the Creative Principle is described by a set of information that not only includes, but circumscribes everything in creation. That's a very large set of information. One word that could summarize that state of consciousness would be *"omnipresence."*

The Creative Principle is a state of consciousness that includes, and is aware of, everything going on in creation simultaneously. The entire operation of the Creative Principle through the Field is not one of matter, but one of consciousness. Each of our physical forms came into existence through the omnipresent consciousness of the Creative Principle, and we are all engaged in the process of remembering who we really are and from whence we came. According to certain wisdom traditions, the Creative Principle exists in a state of bliss that is so magnificent it cannot be captured by the use of words. Hindu sages teach techniques in the use of free will that expand the devotee's consciousness to the point where this state of ecstasy can be experienced here on Earth. By definition as a minute division of the source consciousness of the Creative Principle, we are all capable of experiencing bliss.

In scientific terms, when the subset of information that is your own subjective universe approaches the grand set of information that is the Creative Principle, you'll shed your identity as a separate consciousness and merge back into the ineffable bliss in which the Creative Principle resides. In some religious traditions, this state of merged existence is referred to as "coming home" or "sitting at the right hand of God in heaven." This exalted state of being doesn't result simply from living one good life. It doesn't result from a simple declaration of belief in a certain individual. It results from doing the sacred work of becoming a living example of truth no matter how many cycles of incarnation are required to accomplish it.

When all of our intentions and actions in daily life originate from the intuitive wisdom of our hearts, when we feel and act with sincere appreciation, caring, and kindness for others, and when we can observe the world around us without the influence of preset judgments of the mind, but rather with compassion in our hearts, then we'll have changed our individual belief system into a cosmology worthy of the 21st century. Then the individual demonstration of the truths that we each present to the world will cause our collective belief systems to change and the world around us will, at last, begin to transform in a way that is inspiring to all.

Chapter 9

Co-Creation and the Law of Unintended Consequences

While technology is advancing all around us we seem unable to use it to improve life for a majority of our planetary citizens. We are embracing new technologies at such a blistering pace that we don't seem to have time to properly assess the unintended consequences of our actions before we realize what they are. By 1955, half of all American households had acquired television. Prior to that time advertising was limited to print and radio. It became clear right away that television advertising was a powerful tool. One commercial after another touted the wonderful benefits of smoking cigarettes, persuading an entire generation that they were less than stylish if they didn't take up the habit. The deaths and disabilities that flowed out of that product are one legendary example of unintended consequences. It would take more than a generation before smoking would be frowned upon by the public at large.

During the last 30 years we have manufactured, sold, and are now broadcasting to billions upon billions of wireless devices: radios, televisions, cell phones, laptops, wireless routers, pagers, cars with satellite connections, garage door openers, and remote controls galore. We are systematically turning up the power of wireless networks of every description without a thorough understanding of the effect electromagnetic broadcasts may have on human biology.

Today, it's quite normal to see people wearing a cell phone device clamped to one ear. Do we really know what the long-term consequences of such developments will be? We are finding that the use of these devices while operating motor vehicles is causing a higher rate of accidents. One response to the problem is to ban the use of cell phones while driving. Another response is to design cars that drive themselves so the occupants can stay connected at all times. Cars that park themselves and cars that apply the brakes without driver input are on the market now. The unintended consequence that could arise from this way of problem solving is; the more foolproof we try to make our environment the more we produce generations of fools to inhabit it. Imagine what happens after a few generations have been riding around in completely computer controlled cars, then the computer system is consumed by a virus and the humans have to take over. The occupants of those cars no longer have the driving skills or reaction times to cope with what the computer has been doing for them all these years. The results could be catastrophic.

Here in America, a very high percentage of the built environment has been put in place since World War II. Most of our development has been designed to accommodate the love affair we're having with the automobile. When I was a boy the family used to go for Sunday drives in the country as a form of entertainment. A half century later we have unprecedented levels of air pollution, thought by some to be causing global climate change. If the news had come over the radio of our 1953 Pontiac Chieftain that our Sunday drive would be causing devastating climate change we wouldn't have understood the connection, much less predicted the consequence.

Our "brilliant" financial sector invented credit default swaps, subprime financing, and mortgage-backed securities without analyzing their potential consequences beyond short-term gain. This whole house of cards was based upon the generally-accepted idea that real estate prices would continue to climb. When real estate prices dropped, the bottom fell out of every deficit-spending economy on the planet. For the most part, this effect was unforeseen.

Unintended consequences are not necessarily all negative. One of the positive outcomes of the global financial crisis is that we are receiving a practical demonstration of what the sages have always known, which is:

The extent to which all humanity is deeply connected.

I remember being profoundly moved upon seeing the first photo of the whole Earth taken by the Apollo astronauts over 40 years ago. Until that photo was released to the public, the Earth seemed so vast we couldn't embrace the idea that humanity is actually aboard a spaceship floating in deep space with limited resources available for our survival. In the same way that it took slightly more than one generation to accept the idea that smoking can be hazardous to our health, it has taken roughly the same amount of time for us to hear the message of whole-Earth consciousness.

That which each one of us chooses to do locally eventually will affect all of humanity globally.

While the current financial crisis is destroying political systems, financial systems, and currencies, it is also demanding that we begin to do a much better job of managing the unintended consequences of our actions for all passengers on spaceship Earth. The concept of unintended consequences was first introduced during the mid-18th century by Adam Smith, a Scottish social philosopher. The idea was popularized by Robert K. Merton in his 1936 paper "The Unanticipated Consequences of Purposive Social Action."[1] More recently the concept has been referred to as the Law of Unintended Consequences. My version of the law reads:

Every creative act you engage in sends an electromagnetic wave of intent out into the Field. The Field accepts your creative act as a statement of "what is" in your universe and produces a manifestation that returns in the opposite direction to you. The manifestation that comes back to you carries a content of emotional energy equivalent to that expressed in the original creative act, along with an unintended consequence.

All acts may be considered creative acts. The resulting appearance of an unintended consequence is a clue indicating that there are parts of your consciousness which are manifesting your reality that you may not be fully aware of. The purpose of the unintended consequence is to make you aware that there are deeper levels to your consciousness yet to be explored.

You are having an ongoing, nonstop conversation with the Creative Principle though you might not be aware that such a thing is happening. Your side of the conversation is the creative input you are offering to the Field. Your creative input is made up of the thoughts, feelings, spoken assertions, and actions you contribute to life in every hour of every day. The other side of the conversation is the reaction to your input that shows up in the form of your life. You may recognize the stage in life where, despite what you think your intentions are, a surprise always seems to accompany everything you try to do. The continuing presence of such surprises might be your first indication that you are participating in a conversation. The purpose of unintended consequences, or "surprises," is to tell you to pay attention, what you need to learn in order to expand your consciousness is contained within this experience.

There are three different types of unintended consequences for even our most intentional creative acts, like building a shopping center or writing a book. These are:

1. A positive unexpected benefit or synchronicity.

2. A negative unexpected detriment.

3. A perverse effect contrary to what was intended.

A positive unexpected benefit or synchronicity is what happens when you meet the right person who will help you achieve your goal, or you show up in just the right place to take advantage of a circumstance that could only happen if you were there. It might be an unexpected monetary windfall or even finding your heart's desire in the love of your life.

A negative unexpected detriment might have occurred when you try to help someone less fortunate than you only to find they have adopted you personally as the solution to all their future problems. It might be what happens when you hire someone who has been recommended to you only to find that you have a severe personality conflict. It could be something as simple as buying the car you have always wanted and discovering that the seats aren't as comfortable as the ones in your old beater.

A perverse effect contrary to what was intended occurred in my life when I started a non-profit corporation to provide low income housing for the economically disadvantaged in the early 1990's. I went out on a financial limb to acquire an apartment project we subsequently remodeled and opened for low income occupancy. After the project was is operation for a while I discovered that the tenants considered any landlord to be the enemy and systematically destroyed unit after unit in the project eventually converting me into a low income person. Definitely the opposite effect from what I had intended.

All of these effects can happen in succession to one person. Since you have the free will to create your future any way you want, none of these experiences are the last word on who you really are. All of these effects are products of your consciousness in conversation with the Creative Principle. They are there to guide you. The intent of the Creative Principle, of creation itself, is to expand consciousness, and the mechanism it has put in place to accomplish this is the most natural and elegant system that exists. The Creative Principle consists only of pure consciousness. Because it has no form of its own, it created the Field, which is not physical. The Field is etheric; it is the background energy from which all form is made. The Field has been created to reflect who you really are back to you so you can understand how the mechanism of life works and evolve into a higher awareness.

Every sentient being can be looked upon as a small, self-aware piece of the Creative Principle. In the same way that many humans produce offspring, you could say that the Creative Principle has given birth to your self-awareness. Once that separation occurs, your individual consciousness is allowed to descend through a succession of more dense worlds, culminating in the physical plane. It's as if your parent has sent you off to college somewhere in outer space and still wants to keep track of you.

The Creative Principle wants to do so by enjoying every sensory input you experience, and it has endowed you with the free will to create any situation you want so that it can experience everything possible within creation. In a sense your biology is an extension of the nervous system of the Creative Principle. The purpose of your biology is to transmit your experience of creation back to the Creative Principle, which expands its self-awareness along with your own.

In the conversation between you and the Creative Principle, the language that is being spoken in the physical plane is the language of feeling. The Law of Unintended Consequences provides that every creative act you engage in will produce an equal and opposite reaction in your feeling state. While your conscious awareness is developing, these feelings almost always come with little attachments from the Field. These unintended consequences are the other side of the dialogue.

An unintended consequence is provided to you by the Creative Principle as a teaching mechanism. It shows up in your life as something that comes seemingly out of the blue. You didn't think of it. It wasn't a part of your plan, but there it is. You could think of the unintended consequence as being the punctuation marks within your conversation.

An unexpected detriment or negative surprise is telling you that you haven't quite mastered the system of manifestation yet. It occurs as a guide to help you learn by rote which types of input create which types of results. A perverse effect contrary to the original intent behind your creative act is a sign that you are probably not fully aware the conversation is being had. You may interpret such experiences as proof that you cannot get what you want in life, but that is an erroneous belief.

An unexpected benefit or positive surprise (which is also known as a synchronicity) shows you that you are beginning to understand how the system works. It means you are beginning to understand that the Field does not react to what you say you are but to how you feel within.

Once you understand that your life is a conversation with the Creative Principle, you can begin to look with longing toward the appearance of the next unintended consequence, knowing that this is how you'll learn. You will see that in the same way your parents don't want to be forgotten by you, the Creative Principle wants you to recognize that you have a relationship with it. The Creative Principle wants to dance with you. It wants you to feel the bliss of ever-expanding consciousness, but allowing for the free will of all possibilities includes the possibility that you won't allow yourself to feel that bliss. You have to provide the effort needed to discover what bliss is.

When you begin to feel a kind of happiness that wells up from within you quite independently from what is happening around you—when negative unexpected consequences don't upset your equanimity—then the other side of the conversation begins to produce synchronicity to show you that you are on the right path. That synchronicity is the evidence that you are beginning to co-create with the Creative Principle intentionally.

The process of co-creation rises to a new level after you have demonstrated to yourself that you actually are having a conversation with the Creative Principle. The process of intentional co-creation really begins once you have willfully created input with the specific intent of recognizing the response of the Creative Principle through the unintended consequences that pop up. When you have made a mental connection between intentions and surprises, and have repeated the cycle consciously a few times, then you are on your way to living a life of synchronicity.

A synchronous life is a fun life. You meet the right people, you find yourself in one situation after another that leads you toward a previously unobtainable state of fulfillment. You no longer need to struggle to stay afloat. Life becomes much easier. All of this is produced from inside of you. On the physical plane most of this is produced by the creative act of what you choose to feel inside. The secret is:

It is not what happens to you, but how you react to what happens to you, that makes the difference in your future experience.

What there is to do is to learn to manage the feeling of your creative input. You don't need to take a course to do this. You don't need a certificate. It doesn't cost anything to begin. You can start right now, this very minute, to change your life forever. And that's what we'll discuss how to do in the next chapter.

Chapter 10

The Emotional Spiral

The human body is a curious evolutionary form. It is festooned with an array of sensors that make the Star Ship Enterprise pale by comparison. We have eyes to see, ears to hear, a nose to smell, a tongue to taste, and skin to feel with. Each one of the organs of sense has millions of cells whose job is to gather information from the world around us and transmit it to the brain where we coordinate this information into an overall picture of our environment.

Of the five sense organs, the skin is by far the largest. Pick a hair anywhere on your body and give it a little pull. You can feel the sensation. Take off your clothes and gently run a feather over every part of your skin. The skin is designed to feel. The other senses are also ways to feel what is present in our environment. Clearly, the human body is designed to feel.

There is another form of feeling that we are all familiar with. Emotion is defined as any of the particular *feelings* that characterize a state of mind, such as joy, anger, love, or hate. The sense organs take in information about our surroundings. They perceive it is hot or cold or bright or loud. It has a smell and a taste to it. We then employ our emotions to decide how we *feel* about the information gathered by our *perceiving* senses. Notice that we commonly use the same word, "feeling," to describe the two different functions of physical sensation and emotional feeling without separating them into distinct operations.

If you just missed another car at high speed you might decide the information you've gathered is life threatening and you might feel fear. If you just tasted the most delicious pie you've ever eaten you might decide the information you have gathered is satisfying and you might feel pleasure. The emotion you feel is a separate and distinct operation from the gathering of sensory information that you engage in. It appears that the event you just experienced is the direct and instantaneous cause of the emotion you are now feeling. That is to say, it appears that the cause of your emotion is something that happened outside of you. Emotional responses usually happen so quickly that we don't realize there is a choice-point between sensing and feeling where a creative act took place. But there is. This creative act is where the feeling is chosen. It is where we can select our emotions and begin to co-create with the Field.

Most of what we've experienced during our lives is repetitive. It is somewhat rare to experience a completely new sensation. Somewhere in your sensory past you decided how you *feel* about a particular kind of input. Whenever that input shows up you defer to the default setting of your previous decision. The sense of having engaged in a creative act by deciding how you feel about that input seems to disappear. The event happens and you feel. The distinction that you actually decided how to feel about the experience is lost, creating an illusion that the event caused the feeling.

While an event does cause a sensory feeling, the truth is that you can choose the emotional feeling that goes with it. Somewhere along the line you may have decided you abhor the use of a particular curse word. Whenever someone uses that word in your presence you allow the same negative emotion to be triggered instantly. Like most people, you are unaware of having made such a decision. It appears that whoever did the cursing is the one who created the emotion you are now feeling inside.

The emotional device of pointing a finger toward another person and saying, "Look how you made me feel," is very common in today's society. By abdicating the responsibility for having made a decision in the matter, the illusion is created that we are the victims of other people. If you feel that you are being swept along by life with no control over how you feel, you have adopted this way of being.

The antidote for feeling you are being swept along out of emotional control is to take the time to go within and observe what you are really doing. The first lesson in this portion of our curriculum is to realize that no matter what sensory data you take in, you have made a decision to feel what you are feeling.

For now, let's not discuss the immediate response of the fight-or-flight mechanism. There are occasions when a life-threatening situation occurs that results in a jolt of adrenaline producing an immediate reaction that can be very useful to you. Those reactive instincts are designed to preserve your physiology and I am not going to recommend that you dismantle them. As we go forward I want you to focus instead on the emotions that make up your overall sense of well being.

Learning How to Manage Your Creative Input

It is estimated that the average person has between 10,000 and 20,000 thoughts per day. For the sake of conversation let's use the figure of 10,000 thoughts per day. Each of those thoughts is accompanied by an internal sense of feeling. A large portion of those feelings are relatively "benign," meaning they have little impact on your manifestations and produce few detrimental or perverse consequences. The rest of those feelings make up the magnet that pulls experience toward you to fulfill your personal declaration of "what is."

Because your consciousness is directly descended from the Creative Principle it has a fraction of the power that created the entire universe within it. When you declare "what is" with your emotions, the Field accepts that information as a command and produces more of it in your life to fulfill its side of the co-creative conversation you are having.

The idea that consciousness remains intact after the death of the physical body suggests that the seat of the thinking function might not be located in the physical brain. If it were, then conscious awareness would cease to exist upon death. It's possible that the seat of the thinking function is located within the more subtle astral body, and the brain is a receiver that brings the thinking function into the physical body. A polygraph, for instance, does not measure your thoughts. It measures the physiological reaction you are having to your thoughts. The thoughts you are having are not really part of the physical plane.

In a similar way, feeling appears to be more the language of the physical plane. Our bodies are covered with sensors so we can feel what goes on in the physical world thereby expanding our body of knowledge and our consciousness. I believe that both thought and feeling manifest the circumstances of life. But because feeling, whether sensory or emotional, is the language of the physical plane, I believe this is the area of life that has the greatest effect on the physical consequences that show up in our world. I also believe that the dividing line between how thought and feeling manifest in life is personal. Some people are more oriented toward feeling while others are more oriented toward the thinking process. Through my own personal experience I have become convinced that what we choose to feel is much more powerful in manifesting what happens in the physical world than what we choose to think.

I'm known as a fairly competent thinker. I am able to envision highly complex projects and carry them through to completion. As a young man, I designed a line of expandable homes, went through all the stages of financing, governmental approval, site-work and construction in order to bring my thoughts into manifestation. The concept was well executed. I received the hearty approval of my investors for a job well done. Yet the project still failed causing me untold personal pain and grief.

Some folks said it was just bad timing. Others claimed it was the result of too high a debt to value ratio. That recession was very selective. Some people did well, others were wiped out. But, I have had years and years to think about why it happened to me. I have concluded that

I had such a strong fear of failure every step of the way there was no possibility that the project would have been financially successful. That fear of failure was an emotion. Once I admitted that to myself, I began to look at all the other unintended consequences that have popped up in my life and I saw the same thing over and over again.

While I was busy being a competent thinker I was not fully aware that fear was what was running the show. Has someone ever looked at you and asked you out of the blue, "What were you just thinking?" Have you ever responded by claiming that you weren't thinking anything significant to hide the fact that you were so lost in thought you weren't even aware on a conscious level of what you were thinking? It's almost like you just awoke from a dream and can't remember what the dream was about. The same thing is true of our feelings. We may not be fully aware of what is going on in our minds as well as in our hearts. If you recognize yourself in any of the preceding paragraphs, you are already on your way to experiencing a higher level of consciousness since the first step is becoming more aware of what you have actually been doing.

Here is how you can learn to manage the feeling of your creative input in the conversation you are having with the Creative Principle:

1. The first step in learning how to manage your creative input is to accept the working premise *that your thoughts and feelings are the input that creates your reality.* I'm not asking for blind faith in a belief system just because I said so. I'm simply asking you to believe in the possibility this is true long enough to

prove it to yourself.

2. The second step is to begin to monitor your thoughts and feelings all day long every day. You may have heard of this practice of being the witness. The more you witness your internal processes of thinking and feeling, the less you will have those sessions of being so lost in thought you aren't aware of what you were actually thinking. This practice will increase your awareness.

   If you do this correctly, you should be able to sit down with your spouse or a friend at the end of the day and recount the highlights of your thinking and feeling during the course of the day. Don't just tell your spouse or confidant what happened, tell him or her what you thought and felt about what happened. That will help to bring your inner processes out into the conversation.

3. The third step is to apprehend the fact that you are *deciding* how to feel about what happens to you and those emotions are not being imposed upon you. Right after an event occurs there is a choice point that precedes the emotion you decide to feel. It happens so quickly the process appears to be seamless. The event happens and you have the feeling. To start with, you will probably see this function in retrospect, after you review whatever event just took place and the emotion that you felt along with it. That's fine. You can still learn the lesson by observing it after the fact. With practice you'll learn to catch yourself in mid-emotion and decide whether or not you want to continue what you are feeling.

   To remind myself to do so, I have a little mantra that I use. It goes like this:

Whatever I am thinking and feeling in this moment, I am asking the Creative Principle to produce more of in my life.

If I am busy visualizing something negative happening I ask, "Why would I want more of that?"

4. After you have practiced the preceding three steps for a while you'll notice that you have repetitive thoughts and feelings about things you don't want more of in your life. When I catch myself in mid-emotion, feeling something I definitely do not want more of, I don't deny what I'm doing and I don't suppress what I'm doing. I change the channel. I think about something I do want more of, and then I try to feel what it feels like to already have it. This is the tricky part. If I think that I don't have enough money, I feel a sense of lack. The Creative Principle will honor my instructions, I have decreed that there is a lack and so lack appears.

A typical thing to do when I don't get what I want would be to turn up the power of my wanting to a higher level. Now I really, really want to have the money I don't have and the result is the same, but more intense: I do not have money.

Wanting is not having. The tricky part of interrupting this pattern therefore is to think, regardless of whatever circumstance surrounds me, "I always have just the right amount of money to do whatever I want to do," and to feel the physical and emotional sensation of always having enough, as if it were already true. Then the Creative Principle will honor my instructions and then the object of my thinking and feeling will come to pass.

Now, managing my emotional input gets even trickier. I am having 10,000 thoughts and feelings every day. A momentary inspiration in which I've created the preform of a prosperous life isn't going to manifest a prosperous life if it is merely one thought in 10,000. That inspiration needs to take hold and become a habitual part of who I am before it happens and despite what is happening to the contrary in my life. Then it will manifest.

If you are used to looking at life from the point of view that

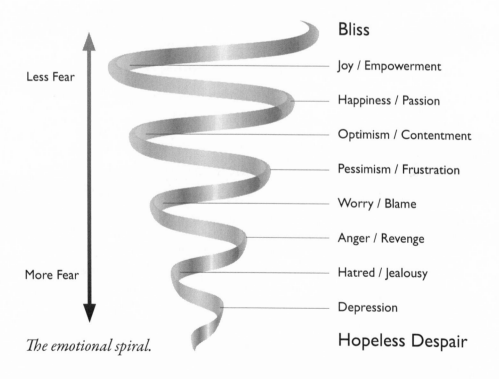

The emotional spiral.

you "never really get what you want," you have a lot of internal work to do. The good news is you have 10,000 opportunities to do it every day.

This process will take longer in some people than in others.

How quickly the change occurs in you depends on how negative a person you have been all your life.

The Emotional Spiral

To illustrate the way emotional input works, take a look at the illustration of the emotional spiral on the previous page. The emotional spiral consists of a hierarchy of emotions ranging from hopeless despair at the bottom to bliss at the top. In the picture, the spiral is tapered to show that the emotions at the bottom are more restrictive and the emotions at the top are more expansive.

We can say that each one of us habitually lives on one of the levels of the emotional spiral as a way of being. Each level represents a theme of the thinking and feeling present in a certain body of knowing. The emotional spiral is a hierarchy because in an overall sense we must pass through each level of emotion in succession to get to the next level. That's not to say we can't jump from Worry/Blame to Happiness/Passion in any given moment. But if we habitually fill our hearts with Worry/Blame we won't be able to stay in the higher emotions, we'll spiral back down.

For our current purpose, let's refer to whatever level you exist on most as your *"emotional baseline."* The goal of expanding your consciousness is to raise your emotional baseline permanently.

In today's world the most common baseline I find people living in is Worry/Blame. Let's face it: There are a lot of things to worry about going on in the world. When there are things to worry about, it is natural to look for who is to blame for the sorry state of affairs we find ourselves in. If you are a worrier, I must caution you. The more you worry about life, the more the Creative Principle accepts this as your instruction and so it will produce more and more things for you to worry about. The more you blame others for the fact that you are worrying, the more people show up and do things to you supporting your decision to blame and you begin to spiral down.

Blame is a kind of dividing line between the upper half and the bottom half of the spiral, because the act of blaming is the act of giving up your power. The natural outcome of a heart filled with worry and blame is that it begins to turn to anger. The angrier you get over your belief that life sucks, the more helpless you feel. The more helpless you feel, the worse your situation becomes. Then the need for revenge shows up. You not only want to blame the perpetrators responsible for your sorry state of affairs, you want to even the score.

If you can't get a satisfying level of revenge (and you never can), then you begin to hate those who have unfairly imposed upon your life. The more you hate, the more things show up to bring out the hate that is within you. What goes hand in hand with that is the jealousy that those who impose upon you always seem to have it easier than you do. You might think, "It is so unfair." The more jealous you become the more there are things to be jealous of. You spiral down and down. Then we come to depression. Nothing ever works. It doesn't matter what you do, everything always turns out the same. Life really sucks. You always get screwed by everybody else.

Finally you bottom out in hopeless despair. There are no solutions; you have tried everything there is to try, except changing what's in your mind and in your heart. When you are trapped in hopeless despair you begin to think about ending it all. Here is the problem with that. Consciousness is eternal. If you pull the trigger, you will find yourself still living the same baseline, just without a physical body.

Can you begin to see how we spiral down?

In presenting this material to lecture audiences I've noticed something very interesting. Most people claim they are living a step or two higher on the emotional spiral than they really are. When you have a few casual conversations about life with someone it's not difficult to pin down their emotional baseline. Time and time again I have run into people who clearly live from a baseline of Worry/Blame saying that they live in Happiness/Passion. It's very common to be deluded about this. You might want to examine the tone of your thoughts for a while before you decide where you are starting.

By now I hope you see that your emotional interpretation of what is happening to you in life is completely subjective. Once you have monitored your thoughts and emotions for a while, you will begin to see that you have more options than simply playing the prerecorded mental and emotional reactions you've been recycling all your life. You might want to play with creating an opposite emotion.

Spiraling up is more difficult than spiraling down because it requires eternal vigilance. Doing nothing to improve your life is much easier than making a concerted effort to expand your consciousness. However, there is a simple way to begin to spiral up. When something happens that you habitually feel bad about, catch yourself in the act and decide on the spot to feel happy that you have caught yourself in the act of playing a pre-recorded emotional message. That very simple act of being happy with knowing where you are now changes everything. With it, you have begun the process of spiraling up.

If you practice this technique for a while you can achieve some skill at it. The catch is to feel better, as an act of creativity, despite the fact that the life showing up around you is still based upon your old baseline. You will actually feel better with no outside reason to do so. You will create that feeling with the power of your own free will.

Then you begin to take your power back. You begin to remember who you really are. Even if you live in Worry/Blame you start to ascend. It's interesting, you may be shocked to see how habitually pessimistic you have been about life after you practice creating a happy reaction to life with no external reason to do so. Doing this for a while could bring you face to face with the fact that you have always felt frustrated by life no matter what happens to you.

Feeling pessimistic or frustrated is a higher way of being than always being worried and blaming others. When you begin to perceive that you've been a pessimist, you'll naturally have spurts of optimism, since you are starting to see how the emotional spiral works. The more you have spurts of optimism, the more life gives you things to be optimistic about. When that starts to happen the more you will feel contentment. Life is starting to be more fun. You are spiraling up.

As you practice your new skills of feeling Optimism/Contentment, you become able to make little jumps higher and higher. As you learn these skills you may still slide back down to a baseline of optimism, but you realize you are miles above Worry/Blame and that is a happy realization. So you jump up to Happiness/Passion. You become passionate about the pursuit of happiness. It's the best thing ever. The more you inhabit that baseline, the more you are momentarily able to access real joy. Joy is contagious.

Real joy is distinct from mere pleasure. Pleasure is born of the physical senses. Real joy is born of the expansion of consciousness. You begin to see that you have the power to affect your destiny. The more joyous you are, the more life delivers events to be joyous about. Synchronicity starts to become ordinary. You meet the right people and your sacred work in life begins to be apparent. The whole universe starts to have meaning for you. Finally, you gain little glimpses of bliss. Bliss is so ineffable words simply cannot convey the experience. You just know you want to go there.

When bliss is your baseline there are still troubles in the world. You still see what is going on. It simply does not disturb you. You are living in the world but you are not of it. Bliss is the highest form of consciousness, and living in Bliss puts you in harmony with the intent of creation.

For years I was puzzled by this quote from the Bible: "To him that hath, all things shall be added. To him that hath not, even what he hath shall be taken away." It never seemed right that whoever hath should get more and whoever hath not should lose everything. When I began to understand the emotional spiral I began to see the truth in this statement. The biblical scribes had left out one word. That word was "joy." They should have written: To him that hath *joy*, all things will be added. (He'll spiral up.) To him that hath not *joy*, even what he hath will be taken away. (He'll spiral down.)

The practices in this chapter are simple to understand and easy to do. When you first start to spiral up, the Creative Principle will meet you halfway in your efforts. Something unforeseen will happen to delight you—an unexpected benefit. Usually it will be something small. When it happens it comes as an intelligence test. If you inhabit a negative baseline, be very careful that you do not write off what happens as if it were a meaningless coincidence. Keep a sharp eye out and when you see a happy synchronicity, give thanks. The Creative Principle is responding to your efforts to start moving up the spiral. Celebrate it. Be grateful that you have expanded your consciousness enough to see it for what it is.

All day every day, to the best of your ability, make a conscious decision to feel better, no matter what is going on around you. As you gain strength and improve your ability to focus, day by day you will start to remember who you really are. You are a powerful creator-in-training and your birthright is the ability to feel joy, and ultimately bliss, continuously.

Chapter 11

Who Am I Now?

The majority of people practicing some sort of religion have a concept that there is life after death. After all, if you must present yourself for the final judgment, there must be some form of your consciousness left to stand and be judged. The people who believe in divine judgment organize their lives in preparation for this event to take place. They have a certainty that there is life after death.

If there is life after the death of the physical body, it shouldn't be a big stretch to conceive that there might be life *before* the physical body is born. This would mean the soul is immortal. In fact, there is only one concept that explains why the Creative Principle allows injustice, genocide, and evil to exist in the world. That concept is the immortal nature of the soul.

The immortal soul manifests a temporary physical body with which to come and experience the physical world as a means to expand its conscious awareness of the principles of creation. If you take a careful look at the lives of your loved ones it is relatively easy to see that the temporary nature of the physical body does not give one enough time to master the laws of creation. I remember my grandmother's last words before she died. She asked the question out loud, "What was I so angry about?" Then she passed from this world.

The experiences the body has on this Earth are not the main event in our existence as an individual unit of self-aware consciousness. The principles behind creation are vast, and it takes most folks way more than one lifetime to complete the curriculum. The idea that we only have one life to live is an illusion. We have already discussed the nature of matter. The atomic structure of the physical universe is not really solid. It is an illusion. So is everything that happens here in the dream of the physical world. Getting a feel for the dream nature of what we think of as reality can take some time. That is not to deny that it really does hurt when we are injured. However, if we weren't allowed to manifest the full range of experiences that occur on this plane, we would not have free will, and the exercise of free will is necessary to building an understanding of how creation actually works. The exercise of free will is necessary to actively and intentionally participate in the process of creation.

The physical world is a world of duality. If there is big, there is little. If there is hot, there is cold. If there is bright, there is dark. We are allowed to come here and thrash around, creating indiscriminately, until we understand what we don't want and eventually, by contrast, what we do want. I spent decades having the same repetitive experience of what I didn't want before I began to wake up and understand what was really going on.

I've been a builder and a developer for over 40 years. The curious thing about these vocations is they are the first in line to be affected whenever there is a downturn in the economy. I've been through five recessions in my career, one roughly every eight years or so. Each time I would work diligently for that period of time, making headway financially, only to see all the progress I made wiped out in a period of a few months by the next down turn. Periodically I spiraled down to the bottom and spent a year or two in mild depression, and some time in hopeless despair. Then I would pull myself up and start over again. I couldn't understand why I was having this repetitive experience. Every time a recession came around, some people in my field were affected and some were not. Nothing seemed to change it for me. I was always one of those who were affected. I hadn't yet figured out that the only thing I needed to change was the contents of my mind and my heart. The only thing I needed to change was my body of knowing.

I had to rattle around in hopeless despair for a while until I was convinced the experiences I was having couldn't possibly be all there was to my life. In order to pull myself back up, I relied on the love I had for my wife (I had married again, this time very happily), the beautiful people my children had become, and the strength I found in nature. I couldn't abandon the intuition that there had to be more to my existence on Earth than I was aware of. I had to seek a higher understanding of the nature of truth and meaning. I had to find some way to put all the experiences of my life into some kind of order. There must a point of view from which it all made sense. I freely admit that I'm a little slow on the uptake. Better late than never.

It's only been a decade since I really started to seek the truth of creation in earnest. Just the simple creative act of deciding there must be a vantage point from which it all made sense changed everything. I became a spiritual private eye investigating life itself. I read everything I could get my hands on that I thought would help me raise my consciousness. I talked to anyone who would listen. One of the hallmarks of being in hopeless despair is the act of looking back at all the experiences one has had in life and coming to the conclusion that none of it makes any sense. I began to see that all is well that ends well. The only thing that changes the past is what you decide to do next.

I began to see that if I could pull myself up, I would be able to look back at all the crazy experiences I'd had and say that each one was a lesson that eventually led to higher awareness. The alternative was to stay groveling at the bottom of the emotional hierarchy, where nothing made sense, and eventually expire having lived a meaningless life.

One of the things that helped me to begin the process of spiraling up from the bottom was the concept that consciousness is eternal. When I was 52, I met a shamanic astrologer who told me we've all come here with original intent. (Shamanic astrology is different from sun sign astrology.) He said, "At 52 years of age you have been through all the cards in the deck and now it is time to change the game." I needed to understand my original intent for incarnating on the planet. If I did so, this would change my concept of where I was going in life. If I got that straight, the story of my life would start to make sense.

He was right about that. Had I been monetarily successful in my early years I would have stopped seeking and rested on my financial laurels. I was that person. I had always placed a high value on monetary success, mostly from the point of view of not having it and being jealous of those who did.

After that reading, I looked at all the tendencies I had displayed during my life. I was good at assembling complex projects. I had always felt a compulsion to be in partnership with people. The shamanic astrologer told me I was a philosopher scout. Each and every one of us has sacred work to do. Everyone's sacred work is different. It was my job to go explore the far reaches of philosophical thought and come back to report to the tribe.

Sacred work represents the fastest path we can each take to reach enlightenment. If I came here and allowed myself to be distracted, if I failed to do the sacred work behind my original intent, somehow life would keep knocking me on the head until I either expired in bewilderment, or remembered what I came here to do. I struggled and struggled. I was a builder and a developer. That's all I knew.

Finally I decided to write down my story. Maybe the act of me telling my own story would help. The more I wrote the more clear it became. The pattern was beginning to be understandable. I was to draw on all the experiences I've had in life, winnow out the cosmic principles, and report back to the tribe. The more I aligned myself with my original intent, the better I felt. The better I felt, the more life presented me with reasons to feel good. Finally, I started to spiral upward past my emotional baseline.

Our consciousness has to exist before we descend far enough into creation to have a physical body. Therefore you were conscious before you came here. In fact you decided to come here to work on a specific aspect of your own enlightenment. You came here with original intent to do sacred work. Incarnating on the physical plane is a conscious act of creation. For most people—not for everyone though—the act of being born in a physical body obscures the memory of a previous existence. We descend into density to fully be here and to fully experience all the feelings the physical plane has to offer.

So how can you uncover your original intent for coming here?

Make a habit out of introspection. Learn to monitor your thoughts. Despite what you have decided upon as a vocation, what's in your life that never changes while everything else swirls around you? Is there some interest you have that you've set aside so you can engage in the rigors of keeping the wolf away from your door? Look there.

For myself, I realized I'd been attracted to philosophy all my life. However, I never considered philosophy as a purpose for being here because I never saw it as a way to make a living. Whatever else I was doing, making a living was always number one on my list of things to pay attention to.

What is it that keeps knocking you on the head? Look there. Understanding what you don't want is an entry point to understanding your heart's desire. That's what we are talking about here: understanding your heart's desire. Part of spiraling up is to be fully engaged in your heart's desire. No matter what you may have done in the past, no matter what errors you have made in life, your destiny sooner or later will be to experience your heart's desire.

You may have mistaken some material object that you don't have yet as your heart's desire. If so, you should concentrate on what you think you'd feel if it had already manifested. What you really want isn't that piece of matter. What you really want is the feeling you think you'll have if you get it. Allow yourself to have that feeling now. The act of feeling is an act of creation. You are a creator. Have the feelings you want without any external reason. Give up the reasons. You don't need them. Learn to acquire the skill to feel what you want to feel no matter what happens out there. Become inspiration itself. When you do that, your sacred work will stand out like a beacon in the darkness. You will spiral up and stay there.

Who am I now? I am a person who is busy learning the cosmic principles that have created the universe itself. Even when trials and tribulations manifest in my life from my old way of being, I am happy and healthy in the knowledge that I am creating a new me 10,000 times a day. I sleep well and feel excited to greet each new day in a state of inspiration for what I might yet become. I constantly ask for and receive more and more of the secrets of creation. The more I master my own happiness, the better life is. I am surrounded by synchronicity. I meet the right people. I arrive at the right time. Truth and meaning are abundant in my life.

I'm not naïve. I understand what's going on in the world around me. I understand that the world we've collectively created in all its dismay and horror is the product of our consciousness. Having achieved a higher level of well being than I once enjoyed, I turn my attention to helping others do the same. I help the inner world change and I help the outer world change at the same time. Who am I now? I am a creator–in–training.

Part III

The Mystery Is Encoded in Nature

Chapter 12

The Precession of the Equinoxes

In Part I we discussed the role of human consciousness in creating our personal lives as well as the evolving world situation. We saw how the Field was created by the Creative Principle in such a way that each one of us is always engaged in manifesting our inner knowing. In Part II we discussed the dynamics of being human, and the creative power we all enjoy as our birthright. Given the advanced state of our best technologies, it would therefore seem that tasks like eliminating world hunger and poverty should be relatively simple to accomplish.

So what is it that holds us back from creating a better world? When we consider the power, intelligence, and elegance that are clearly evident in the cosmos, the gap between our abilities as human creators and the qualities of the world we have manifested makes me wonder if there isn't some unknown force in play that has been restricting our potential. It's not unusual to hear somebody preface a commentary on the state of the world with the question: "How can that still be happening in this day and age?"

Popular culture doesn't yet provide us with much background on any concept of "ages," except to slap the tag of "new age this" or "new age that" on certain publications and music. There is an inkling of an idea in our culture that we might be on the cusp of some sort of age in which things could be improving over all. But have we entered, or are we about to enter, some new age in our solar system that could encourage us to believe that humanity might be able to make a quantum leap into a better future? If we are involved in a cosmic system created with the specific intent of increasing our individual awareness of the whole, it would be nice to locate where we are within that process of realization. Where do we turn for the information? I've said that the answers to the mysteries of life are encoded in the natural world. If that's true, then we should be able to turn to nature for some of the information we seek.

There is a phenomenon that occurs in the night sky that hasn't been well understood in our modern culture. It is called the *precession of the equinoxes.* To demonstrate this phenomenon, let's go back out into our Montana-sized backyard and erect a pair of standing stones. Imagine we've previously made those stones about six inches by six inches in cross section, one being six-feet long and the other about five-feet long. Close to the end of each stone we've bored an eyehole at a slight angle. Now we will sink these stones into the ground about five feet from each other and deep enough so they won't move from year to year. We'll sink them in the ground in such a way that the sightline created between the two holes is pointing at one of the stars in one of the constellations of the zodiac, say the first point in Sagittarius. The shorter one will be our eyepiece and the taller one will be our front sight.

Every year we go out during the evening of the vernal equinox, toward the end of March, sit on a stool, look through our eyehole, and note the position of the star with which we started. If we are relatively precise in our observations we will note that in about 72 years the position indicated by our sightline has precessed (that is, gone backwards with regard to the direction of rotation that the constellations take across the night sky) by about one degree. A simple calculation of multiplying 72 years times 360 degrees gives us the number of 25,920 years for our sightline to make one full backward revolution through all of the constellations of the zodiac.

The ancients, who thankfully didn't have TV and spent their evenings out of doors looking at the night sky, were well aware of the phenomenon of precession. There are many sites around the world where one can find standing stones aligned in the way I've just described, showing that precession was being measured by our ancestors. In my view, one of the great tragedies of the current age is that a

A galaxy like ours.

large percentage of the world's population has never spent any time contemplating the night sky and sees no significance in what may be going on out there.

For a long time the conventional wisdom was that our solar system floats in formation with billions of other stars in a precision orbit that takes about 240 million years to make one revolution around our galactic center. The fact that the stars within our visual range maintain their positions relative to each other would seem to corroborate the idea that they are all moving in formation. But then there is the pesky precession of the equinoxes, which appears to indicate that there might be some additional intermediate level of movement in play. A few different theories aim to explain why we see a precession of the equinoxes in the night sky. One theory is that a wobble occurs in the alignment of the Earth's axis over time. Another interpretation is of particular interest.

Since 2001, The Binary Research Institute has been studying the theory that our solar system is one of a pair of similar star systems locked in a binary orbit. Take a minute and Google "The Binary Research Institute." When the home page comes up, click on "Introduction" at the top and you will see an animation of two stars in a binary orbit, along with detailed material on Binary Companion Theory. It is estimated that as many as half of the stars in the Milky Way Galaxy (our galaxy) belong to binary star systems, so it wouldn't be a big stretch to find out that our own solar system is included in that group.

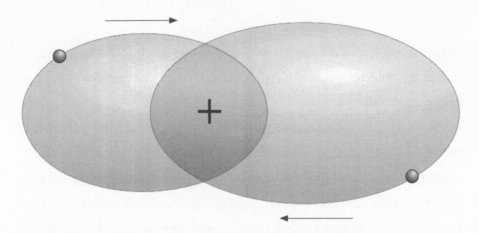

The movement of two star systems around a common center of mass is coordinated by the force of gravity. Each star system occupies the same position in its respective orbit in a mirror image of the other star.

A binary orbit.

I'll leave it to you to explore the BRI website, where you'll notice that the Institute makes a strong case that the precession of the equinoxes is not the result of a wobble in the Earth's axis, but rather the result of our solar system being locked in such an orbit. Each star system moves along its orbit in a mirror image of the other star system. As they move, both stars maintain identical and coordinated positions within, but on opposite sides of, their respective orbits.

The BRI animation of the binary orbit also clearly shows that as two star systems approach one another they accelerate toward each other along the lines of their respective orbits. The increase in velocity that occurs when two celestial bodies of large mass approach each other has been described and verified by Kepler's second law of motion.[1] How would this knowledge affect our calculation of the 26,000 years it takes to make one pass through the constellations of the zodiac?

When we looked through the eyeholes of our standing stones and noted the backward movement of the equinoxes, we assumed that the rate of movement we observed was constant and just multiplied 72 years times 360 degrees giving us the 26,000-year period. But what if the rate of precession isn't a constant? Kepler's second law dictates that there will be a component of acceleration in the equation as two bodies of large gravitational mass approach each other. As a result, we might suspect the actual cycle to be a little less than 26,000 years.

Now that we have seen and measured precession in the night sky, where do we turn for the answer to how long the cycle of precession actually lasts and what its effect on humanity might be? There are only a few people in the astronomical community who have embraced the concept of binary star systems and are actively looking for a companion star to our sun. So far it hasn't been identified by science, but that may be because the distances are long and the calculations arduous. However, having not yet identified a companion star, we can't use it to make direct observations.

Is there anywhere else we can turn for the answer? Is there anything available in ancient lore that could shed light on the subject? As it turns out there is. In 1794, a very old Sanskrit document known as *The Laws of Manu* was first translated into English by Sir William Jones, a British scholar of ancient India. In India, Manu is considered to be the progenitor of mankind, a figure not unlike the Christian concept of Adam. In the translation of this document, Manu refers to an orbital period of approximately 24,000 years, which is divided into progressive ages of enlightenment. In 1894, 100 years after Jones' initial efforts to verify this information, the subject was taken up again by the illumined master Swami Sri Yukteswar of Serampore, Bengal.

Sri Yukteswar was the spiritual teacher of Paramahansa Yogananda, the founder of The Self-Realization Fellowship (SRF) headquartered in Los Angeles. Yogananda's book, *Autobiography of a Yogi* (Self-Realization Fellowship, 1946), provides a detailed description of the swami's life and abilities, including his resurrection after death.[2] Sri Yukteswar's own book, *The Holy Science* (Self-Realization Fellowship, 1949), is a difficult read, but in my view represents one of the highest authorities not only on the 24,000-year cycle, but also on the nature of reality and humanity's role in it. That having been said, everything that follows on this subject has passed through the filter of my limited perception and does not necessarily represent the views of Swami Sri Yukteswar or SRF except where direct quotes are used. I bring this material to your attention because I think there is a message here that we might find useful in crafting our cosmology for the 21st century.

Here is a quote from *The Holy Science:* "Moons revolve around their planets, and planets turning on their axes revolve with their moons around the sun; and the sun, with its planets and their moons, *takes some star for its dual* and revolves around it in about 24,000 years of our earth, a celestial phenomenon which causes the backward movement of the equinoctial points around the zodiac."

In essence, in this passage we find a saint who conquered death claiming that our solar system is part of a binary star system, and that this binary orbit is responsible for the precession of the equinoxes in a 24,000-year cycle. Yukteswar then tells us: "The sun also has another motion by which it revolves round a grand center called *Vishnunabhi*, which is the seat of the creative power, *Brahma*, the universal magnetism. *Brahma* regulates dharma, the mental virtue of the internal world."

These two passages describe both a binary orbit of our solar system around a common center of mass shared by another companion star, and an orbit that the whole binary assembly might take in rotating around the center of our galaxy.

It should be noted here the conventional interpretation that our sun is simply free floating in outer space along with more than 200 billion other stars in a mass orbit around a galactic center does not concur with Yukteswar's wisdom on the subject. The Swami specifically says that our sun is part of a binary star system. I sincerely believe that, once verified, the implications of binary companion theory will be a milestone in our investigations of astronomical structure equal in magnitude to Copernicus' realization that the Earth indeed revolves around the Sun.

One would suppose that an enlightened master would understand the structure of our galaxy and Yukteswar's writings indicate that he did. Perhaps it would have been futile to put the matter before the general public at the end of the 19th century. The first picture of the spiral nebula Andromeda was taken in England in 1885. By 1894, astronomers had not yet realized that our galaxy shared the same structure as Andromeda and had a luminous galactic center. It would take until 1990 before we could launch the Hubble Space Telescope and really observe what is present in deep space for the first time. Through the capabilities of the Hubble program we have made a very quiet quantum leap in our knowledge of interstellar space over the last 20 years.

What Yukteswar does say, is that the grand center is the source of the creative power *Brahma*, the universal magnetism. Sound familiar? The creative power, the universal magnetism, radiating out from galactic center appears to be none other than our previously described electromagnetic Field of All Possibilities. What I am suggesting to you here is that the Field emanates from galactic center like an etheric wind. That wind, consisting of subatomic waves of pure potentiality, varies in its intensity depending on how close one is to the center of the galaxy. Its intensity regulates the creative power of human consciousness.

The study of planetary magnetic fields suggests that the etheric wind of the Field does not emanate from galactic center in the same concentric way that a wave generated by a pebble dropped in a pond does. Such studies suggest that the Field emanates from galactic center in a spiral shape similar in form to the galaxy itself.

The motion of our solar system through space takes us closer to galactic center for a period of time during which there is acceleration in the development of human awareness. The motion of our solar system through space then takes us away from galactic center for a period of time during which there is a corresponding decrease in human awareness. What we will begin to see as we investigate this concept further is that our solar system is entering a region of galactic space where the resulting manifestations of consciousness are coming to their natural conclusion much quicker than we've been accustomed to in the past. The process of manifestation is speeding up.

Virtually all planetary orbits are elliptical. Some, like the Earth's orbit around the Sun, deviate from the circular by an eccentricity of only 2 percent, which is one reason our oceans don't periodically freeze solid or boil. Other orbits are far more eccentric, meaning that they are longer and skinnier ellipses. Our binary orbit through space appears to be much more eccentric than our orbit around the Sun.

In *The Holy Science*, Yukteswar divides our binary orbit into two 12,000-year periods. During one of these periods our solar system approaches its dual star system and, at the same time, approaches the galactic center. Because during this period of the orbit we are ap-

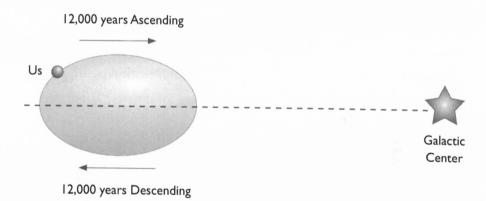

12,000 years Ascending

Us

Galactic
Center

12,000 years Descending

proaching the local source of the Field (the grand center) it is referred to as the "ascending cycle." For the remaining 12,000-year period, our solar system is traveling away from the local source of the Field and is referred to as being in the "descending cycle."

Ascending and descending cycles.

Yukteswar also writes: "When the sun in its revolution round its dual comes to the place nearest to this grand center, *dharma*, the mental virtue, becomes so much developed that man can easily comprehend all, even the mysteries of Spirit. After 12,000 years, when the sun goes to the place in its orbit which is farthest from *Brahma*, the grand center, *dharma*, the mental virtue, comes to such a reduced state that man cannot grasp anything beyond the gross material creation."

This passage tells us there are seasons in the development of human consciousness and that each different season has a specific beginning and ending as a result of the motion of our solar system through space relative to galactic center. Could the current season in human consciousness be the reason we haven't yet reached the implied potential of our role as creators-in-training? If this were true, then what we'd most like to know is: Where are we on our binary orbital path? Where is humanity in the scheme of rising and falling awareness? What can we expect to undergo in the immediate future?

These are complex questions. For the answers we will need to consult the writings of Manu and Sri Yukteswar in more detail. Once this material is digested, we will have a solid sense of our position in the curriculum of the ages.

Chapter 13

The Yugas

According to Manu, each 12,000-year arc of our orbit, meaning both the ascending cycle and descending cycle, can be divided into four distinct ages called *"yugas."* As presented in this book, commencing below, the names of the ages are given in Sanskrit. They could as easily have been named the Golden Age, Silver Age, Bronze Age, and Iron Age. However, since the concepts we are outlining will be foreign to most Westerners, we might as well go ahead and learn the original Sanskrit names for the ages.

In his *Laws*, Manu left us the following description of the ages: "Four thousands of years is the length of Satya Yuga, the golden age of the world, its morning twilight has just as many hundreds and its period of evening dusk is of the same length. In the other three ages, with their morning and evening twilights the thousands and the hundreds decrease by one."

You will note from the preceding description and the following chart that each age decreases in length from the Satya Yuga through the Kali Yuga. Each age is preceded by a period known as a "morning twilight" and ends with a period called "evening dusk." In his commentary on Manu's laws Yukteswar refers to these elements as periods of mutation.

The Satya Yuga is $\quad$ 400 + 4000 + 400 = 4,800 years
The Treta Yuga is $\quad$ 300 + 3000 + 300 = 3,600 years
The DwaparaYuga is $\quad$ 200 + 2000 + 200 = 2,400 years
The Kali Yuga is $\quad$ 100 + 1000 + 100 = 1,200 years

One age of the gods = 12,000 years in length

According to Manu, our solar system experiences a 12,000-year period when it recedes from galactic center, during which all knowledge of the Creative Principle is gradually forgotten. Rounding the point in its orbit that is furthest from the center of our galaxy, our solar system begins to accelerate into a 12,000-year period approaching galactic center.

The dawn of a new age of the gods brings an increase in the intensity of the Field of All Possibilities. Humankind travels through the four stages in the development of human understanding from the Kali Yuga to the Satya Yuga. At the height of Satya Yuga, the Golden Age of the world, the human intellect is once again fully developed and comprehends all, even the Creative Principle known as God, the spirit behind our physical world.

Yukteswar gives the following description of each of the ages that occur during the ascending arc of the Solar System's orbit. "The time of 1,200 years during which the sun passes through a 1/20th portion of its orbit is called the Kali Yuga. *Dharma,* the mental virtue, is then in its first stage and is only a quarter developed; the human intellect cannot comprehend anything beyond the gross material of this ever-changing world.

"The period of 2,400 years during which the sun passes through the 2/20th portion of its orbit is called Dwapara Yuga. *Dharma,* the mental virtue, is then in the second stage of development and is but half complete; the human intellect can then comprehend the fine matters or electricities and their attributes which are the creating principles of the external world.

"The period of 3,600 years during which the sun passes through the 3/20th portion of its orbit is called Treta Yuga. *Dharma,* the mental virtue, is then in the third stage; the human intellect becomes able to comprehend the divine magnetism, the source of all electrical forces on which the creation depends for its existence.

"The period of 4,800 years during which the sun passes through the remaining 4/20th portion of its orbit is called Satya Yuga. *Dharma,* the mental virtue, is then in its fourth stage and completes its full development; the human intellect can comprehend all, even God the Spirit beyond this visible world."

Remember, *The Holy Science* was written in 1894. In it Yukteswar diagrams the ages in the shape of a circle. I suspect he did this to provide a simple illustration of the ratio of the lengths of each age. Since a binary orbit is made up of two ellipses with a common focus, however, we can conclude that he wasn't trying to show the configuration of the actual orbit, which would have to be elliptical.

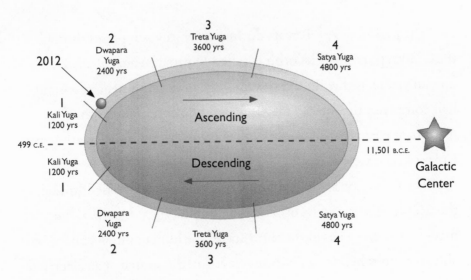

The four ages of our binary orbit.

In the above illustration, I have plotted the orbit according to the ratio of the length of each age. As we proceed we will see that recorded history fits this explanation pretty well. According to Yukteswar's calculations, Planet Earth rounded the point furthest from the *grand center* (another name for galactic center) in the year 499 C.E. The year 499 represents the year zero in our calculations. If we round up for the year one to 500 C.E. and subtract 500 years from the current date of 2,012, the number that remains is 1,512. So our current position in this binary orbit is 1,512 years into a 12,000-year period of acceleration toward the peak of Satya Yuga, the golden age in the affairs of humanity.

Despite the fact that we have some extreme challenges before us, this position in our orbit indicates that overall we should have a lot for which to be hopeful. What could life be like during the golden age of the world? If this whole affair is cyclic in its nature and we have gone

The Creative Principle

through it before, where is the evidence of the highly developed civilizations that must have occupied the planet during the last Satya Yuga? These are intriguing questions. Let's see if we can build a sense of the answers to them.

The illustration of the four ages shows the current position of our solar system in its binary orbit. On the left-hand side of the ellipse is the Kali Yuga, the dark age of the world. This is the part of our elliptical orbit where we rounded the furthest point from our binary companion star. Since then we reached the end of our gravitational tether and are now being pulled back toward our companion star system at the same time as we are being pulled back toward galactic center. We could think of this part of the orbit as a time during which we are pulling a lot of G's, just as we would if we were flying an aerobatic airplane.

During Kali Yuga we might expect to see an increase in the force of gravity, a concept that can be verified by studying the smaller size of human remains and artifacts from the period. We might expect to experience a slowing down in the passage of time, a much more difficult phenomenon to measure than height. Who knows, even the speed of light could be called into question during this period. It was supposedly the darkest time in human existence, a fact that is born out by the history of the period. In fact, we are still trying to escape from the lingering influences of the Dark Ages. As we rounded that tightest radius, the path of our orbit started to flatten out and we began to accelerate back toward our binary companion star, as well as galactic center.

If our theory is to hold water then we should start to see some changes in human affairs that would have occurred during the peri-

135

ods of mutation at the end of the Kali Yuga and the beginning of the Dwapara Yuga. The 100-year period of the evening dusk of the Kali Yuga and the 200-year period of morning twilight of the Dwapara Yuga occurred between the dates of 1599 and 1899, which are clearly marked by the peak of the Renaissance, the beginning of the European seagoing explorations of the so-called New World, and major advances in science and technology.

As we left the tight radius of the Kali Yuga behind and began to embrace the gravitational pull of our companion star, we entered the flat portion of the orbit that is labeled Dwapara Yuga and Treta Yuga. As we started to "go with the flow" of our gravitational attraction, we should have experienced a slight decrease in the force of gravity resulting in an increase in human stature. I am well aware that we chalk up the fact that our children, and especially our grandchildren, are bigger than us to improved nutrition. However, there is cause to be suspicious as to whether the food we eat in this day and age is really superior to the food consumed by our ancestors. Just take a minute and read the labels on what we identify as food in this modern age! It's possible that other factors have contributed to the increase in human stature that we see around us.

Since our location on the flat part of the orbit is where acceleration occurs, we should expect to experience a sensation that time is passing much quicker than before. We're all familiar with that sensation. You can't have much of a casual conversation before someone remarks that they can't believe how fast time is going by. We joke about that concept, but it's very real. Finally, if our theory is to hold water, then during this period of our orbit, we should begin to see a corresponding increase in technology, which is one of the measuring sticks we use to evaluate the state of human consciousness. You would have to be living on a desert island not to be aware of the relentless march of new, more advanced technologies that cross our paths on a daily basis.

Han Solo is about to punch the hyperspace button producing a heretofore unknown rate of acceleration in the affairs of humanity. If you are sitting on top of a rocket it's always a good idea to aim before you push the "go" button. In this case, the aiming procedure would involve taking a look at where we've been, to see where our trajectory might be headed.

Chapter 14

Recorded History

Rather than just accepting what is written in ancient scripture, I felt compelled to examine the material written by both Manu and Yukteswar in more detail, to see how the cycle of the yugas resulting from a binary orbit would overlap with recorded history. The story I've been telling is based upon the premise that the time we spend here on Earth is not necessarily the main event for our souls, but part of a larger cycle of existence. If that is true, then it would seem to follow that such a cycle would have a beginning and an end.

Assuming we are multi-dimensional beings with a purpose, do we descend from the astral world into embodiment in the physical world repeatedly over a 24,000-year period to experience a series of courses on what it means to be human here on Earth? If so, could Earth be a school that we move through on our way to higher and higher expressions of awareness? Admittedly these hypotheses are not something you would find being discussed over a Bud Light at your local tavern. However, in my view, hypotheses that do not stretch the boundaries of what we think of as reality aren't worth posing!

Since, as Yukteswar says, we crossed the point in our orbit furthest from galactic center on a date that falls during recorded history, I decided to look at what was occurring on Earth during that period to see if I could pin down a point in time that would qualify as the beginning and ending of the larger cycle. According to Sri Yukteswar's calculations, the beginning of a 12,000-year ascending cycle occurred during the year 499 C.E. That date falls on the dividing line between the end of the Kali Yuga of the descending cycle and the beginning of the Kali Yuga of the ascending cycle. The Roman Empire had collapsed only 23 years before, as the vandals and Visigoths descended upon Rome from Northern Europe. The affairs of men fell into chaos and pestilence. Ruled only by physical might, this was truly the darkest of times for humanity. While people's lives were turbulent during this period, there is no reference to a single event of cosmic proportions that might signal the beginning or the end of one cycle of the overall orbit.

The next most obvious place to look for what we're seeking—the year that could qualify as the beginning and ending of the whole 24,000-year cycle—is at the opposite end of the orbit, a position that Earth passed through some 12,000 years before. The year 11,501 B.C.E. turns out to have a characteristic that indicates it might be the right date for the end of one overall cycle and the beginning of the next.

An interesting characteristic of the period around 11,501 is that it corresponds with the end of the most recent ice age. At that time, the Arctic and Antarctic ice caps were much larger than they are today. In the Northern Hemisphere, the ice cap covered most of Canada, parts of Washington, Montana, the Great Lakes region, and New England as well as similar latitudes across Europe and Asia. The same phenomenon was present in the Southern Hemisphere, forcing the people of ancient cultures to occupy a much smaller area of the planet than we currently inhabit. The extent of the ice sheets and the corresponding cold conditions that occurred in the higher elevations of all mountain ranges would have forced most civilizations to occupy locations closer to the coasts of each continent.

We know that over 40 percent of the world's population presently lives within 60 miles of the world's coastlines. If highly-developed civilizations existed on the planet in 11,501 B.C.E, we could assume that a substantial amount of that development would have been placed close to the coasts at a much lower elevation than the mean sea level of today. It has been estimated that the melting of the ice caps would have caused a rise in ocean levels from 300 to 400 feet to present-day levels. Therefore, much of what was dry land at the time would have become submerged as the ice melted.

If significant evidence of ancient civilizations exists beyond what we have found above sea level in the current era, we may expect to find it preserved under water a significant distance from our existing coastlines. By using computer models of the extent and thickness of the ice sheets, researchers have been able to recreate the rate at which the ice melted and the rate of rise in mean sea level over the course of history. Once we understand that data, we should be able to calculate the relative age of any megalithic marine architecture we find relative to its depth and distance from existing shorelines.

In his book *Underworld* (Three Rivers Press, 2003), Graham Hancock tells the story of his quest around the world in search of just such evidence of underwater architecture. Hancock documents a number of underwater sites showing evidence of significant megalithic architecture in Cuba, Japan, and Egypt, and along the west coast of

India, dating as old as 7,500 B.C.E.

It is clear that the type of sophisticated architecture that has been discovered didn't pop up over night and is evidence of civilizations that originated prior to that date. If the melting of the ice caps and the rise of ocean levels occurred over a short period of time, this easily could be the event described in the ancient deluge myths told in almost every contemporary culture. In my mind, a worldwide deluge qualifies as the kind of cosmic event that might herald the end of one grand cycle and the beginning of the next. Taking these ideas into consideration, I decided to plot the highlights of recorded history against a portion of the cycle of the yugas to see if they would line up.

Timeline of the Ages: The Descending Arc

11,501 B.C.E.	**Beginning of the descending Satya Yuga**
	End of the last ice age
	Rising of sea level due to melting ice (possible source of deluge legends)
	Disappearance of lost civilizations
7,500 B.C.E.	Sophisticated megalithic architecture built (now underwater)
6,700 B.C.E.	**Beginning of the descending Treta Yuga**
4,000 B.C.E.	Pictographic writing
3,500 B.C.E.	Bronze invented in Thailand
3,101 B.C.E.	**Beginning of the descending Dwapara Yuga**
3,100 B.C.E.	Egyptian hieroglyphics
3,000 B.C.E.	Pyramids built in Egypt

2,100 B.C.E.	Babylonian calendar enters use
1,600 B.C.E.	Temple at Karnak, Egypt, built
1,500 B.C.E.	Phonetic spelling invented
1,400 B.C.E.	Temple at Luxor, Egypt, built
900 B.C.E.	Iron appears in India
701 B.C.E.	**Beginning of the descending Kali Yuga**
680 B.C.E.	Rise of ancient Greek culture
600 B.C.E.	Pythagoras discovers laws of right triangle
563 B.C.E.	Birth of Gautama Siddhartha (the Great Buddha)
300 B.C.E.	Democritus discovers atomic theory
200 B.C.E.	Phoenicians circumnavigate the Earth by ship
100 B.C.E.	Rise of the Roman Empire
47 B.C.E.	Burning of the Library of Alexandria in Egypt
30 B.C.E.	Decline of the Egyptian Empire
0 C.E.	Birth of Jesus Christ
33 C.E.	Beginning of the Christian Church
476 C.E.	Fall of the Roman Empire
499 C.E.	End of the descending Kali Yuga

Timeline of the Ages: The Ascending Arc

499 C.E.	**Beginning of the ascending Kali Yuga**
700 C.E.	Moors introduce Arabic numerals and invent algebra
875 C.E.	Persians apply chemical knowledge to medicine
1000 C.E.	Erikson sails from Greenland to North America

	Gunpowder developed in China
1105 C.E.	First windmill in Europe
1118 C.E.	First cannons used by Moors
1230 C.E.	Hot air balloons invented in China
1260 C.E.	Magnetic compass invented
	Compound lenses invented
1280 C.E.	Spectacles invented
1290 C.E.	Spinning wheel invented
1320 C.E.	Water-driven blast furnace invented
	Beginning of the Renaissance in Italy
1400 C.E.	Diving suit invented
1440 C.E.	Printing press invented
1452 C.E.	Birth of Leonardo DaVinci
1473 C.E.	Birth of Nicolaus Copernicus
1475 C.E.	Birth of Michelangelo Buonarroti
1575 C.E.	Square-rigged sailing ship invented
1599 C.E.	**Transition out of the Kali Yuga**
1605 C.E.	Johannes Kepler develops his laws of planetary motion
1608 C.E.	Hans Lippershay invents the telescope
1661 C.E.	Paper money appears in Europe
1677 C.E.	Isaac Newton discovers calculus
1687 C.E.	Newton publishes Principia
1699 C.E.	**Transition into the Dwapara Yuga**
1708 C.E.	Sand casting of iron invented
1714 C.E.	Daniel Fahrenheit invents mercury thermometer

1720 C.E.	Three-color copperplate printing
1727 C.E.	Measurement of blood pressure
1733 C.E.	John Kay invents flying shuttle, which improves commercial weaving
1738 C.E.	Daniel Bernoulli proves the kinetic theory of gas
1758 C.E.	Achromatic telescope invented
1765 C.E.	Steam engine invented
1770 C.E.	Screw cutting lathe invented
1776 C.E.	U.S. Declaration of Independence signed
1780 C.E.	Mass production introduced
1784 C.E.	Power loom invented
1787 C.E.	Steam boat invented
1791 C.E.	Gas engine invented
1793 C.E.	Cotton gin developed
1796 C.E.	Discovery of photosynthesis
1801 C.E.	First practical submarine built
1808 C.E.	Discovery of sodium, potassium, barium, and calcium
1814 C.E.	Steam locomotive invented
1817 C.E.	Asphalt roads paved in England
1819 C.E.	First steamship crosses the Atlantic Ocean
1827 C.E.	First photograph made
	Discovery of aluminum
1828 C.E.	Electromagnet invented
	Blast furnace invented

1830 C.E.	Sewing machine invented
1832 C.E.	Michael Faraday discovers principles of induction
1833 C.E.	Cyrus McCormick invents reaper
1836 C.E.	Samuel Colt invents revolver
1837 C.E.	Screw propeller created
1840 C.E.	Microphotography developed
1842 C.E.	Doppler effect discovered
	Super phosphate fertilizer developed
1844 C.E.	First telegraph line erected from Washington, D.C., to Baltimore
1845 C.E.	Portland cement developed
1846 C.E.	Discovery of Neptune
1851 C.E.	First cast iron-framed building built
1853 C.E.	Mass production watches invented
1856 C.E.	Henry Bessemer invents steel
1857 C.E.	Elisha Otis invents passenger elevator
1860 C.E.	Dynamo invented
1861 C.E.	Milling machine invented
1862 C.E.	The Monitor battleship built
1865 C.E.	Joseph Lister develops antiseptic surgery
1866 C.E.	Alfred Nobel invents dynamite
1869 C.E.	Transcontinental railroad built in the United States
	Periodic Table of the Elements created by Dmitri Mendeleyev
1873 C.E.	Typewriter invented
1875 C.E.	Telephone invented
1877 C.E.	Phonograph invented

1878 C.E.	Cathode ray tube invented
1880 C.E.	Ball bearings developed
1882 C.E.	Hydro-electric plant invented
1900 C.E.	**Beginning of the ascending Dwapara Yuga**

The preceding timeline does not show every significant development generated by mankind. But it does show a massive and consistent acceleration in human invention and knowledge after the date of 499 C.E. After 1900, new technology starts to be implemented so fast, it would take ten pages to list it all. So here are just a few highlights.

The first automobiles appear

Propeller-driven airplane created

The Federal Reserve System instituted

Massive acceleration in the development of weaponry

Massive acceleration in all forms of electronic communication

The dawn of the Atomic Age

Jet airplane travel becomes commonplace

The thirty-year fully amortized mortgage

The Earth surrounded by satellites

Global Positioning System put in place

Credit cards become commonplace

Men go to the Moon

The space shuttle establishes a presence in low Earth orbit

Computers become commonplace

The Internet revolutionizes access to information

Quantum physics developed

Super string theory developed

Remote controlled flyby and photography of the planets

Massive acceleration in medical techniques

Energy medicine invented

The Hubble Telescope launched into space

The Large Hadron Collider opens for research

That all of these events have occurred in the last 100 years is evidence there is a substantial acceleration in the development of human knowledge taking place. The same acceleration predicted by our
position in the binary orbit. Could this acceleration in the development of human consciousness be a product of an intellectual critical
mass, generated by a growing population more fully applying all of the
information ever generated by humanity? Definitely. Could the speed
of development also be a product of a cyclic increase in the power of
the Field? Yes, it could.

Yukteswar describes the descending Satya Yuga (11,501–6,701
B.C.E.) in a way that suggests this might be the time period from
which many of the ancient myths that have been passed down over
the centuries were generated. It would have been the time when
"gods" walked the Earth and performed astonishing feats. Admittedly,
during the descending Treta Yuga (6,701–3,101 B.C.E.) the timeline is
a little weak. However, quite a few researchers are now suggesting that
the earliest Egyptian civilizations and artifacts are much older than we
think and may have their roots in this period of history.

It's clear that the ancient Egyptians built in a way we would have great difficulty duplicating today. Even if we can accept that the pyramids were built around 3,000 B.C.E., that date puts their construction at 3,499 years before the end of the descending cycle. Since we are only 1,511 years into the ascending cycle at present, the peak of Egyptian culture occurred in a more advanced age, by at least 2,000 years, than the one we currently inhabit. By more advanced, I mean closer to the height of the Satya Yuga, though still within the descending cycle.

Sri Yukteswar mentions that every 12,000-year period brings a complete change. He does not specify what that change might be. One thing we can clearly see in the time period after 499 C.E. is our world culture becoming more mechanistic than anything we see in ancient history. There is very little evidence of the kind of mechanical devices we enjoy today present in the archeological record. We have come to view the level of mechanistic development we live with as the height of sophistication. The ancient Egyptians demonstrated that they could build structures, like the Great Pyramid, which we cannot duplicate today, while leaving no trace as to exactly how they did it. One could easily wonder if they used some sort of technology that is not yet available to us.

I'm not attempting to provide proof that Yukteswar's theory is the gospel truth. However, my studies of both celestial and terrestrial systems have clearly shown me that they tend to be cyclic rather than linear in nature. For no other reason than to promote hope for the continued evolution of our species, I like the idea that humanity may have once enjoyed a higher level of understanding than it does now, and that it may do so again. According to Yukteswar's theory we have passed through the darkest age and are now in the beginning of a period of massive acceleration in the development of the human intellect. If that's true, why are the affairs of humanity still in such a state of chaos?

Chapter 15

The Good News and the Bad News

According to Sri Yukteswar's calculations, by year 2012 we were 312 years into the Dwapara Yuga. Having recently graduated from the Dark Ages, *dharma*, the mental virtue, is in the second stage of development within the ascending cycle. Why then, in an age when consciousness is supposedly advancing, do we find ourselves in such global turmoil? As we accelerate along the path of our binary orbit, shouldn't we be experiencing the steady growth of positive cultural developments resulting from an overall increase in human awareness? If the intensity of the Field increases as we approach galactic center, shouldn't life just keep getting better and better as we speed on our way toward the center of the galaxy? The answers to these questions have several layers.

For the first component, we need to go out into our backyard and once again observe the night sky. Those who live in areas with air pollution and a lot of ambient light sources may need to go somewhere else to see this. Most folks are at least a little bit aware of the path that the Sun takes across the sky. They know that it rises in the East, proceeds across the sky, and then sets in the West. People living closer to the equator have a subliminal awareness that the path the Sun takes across the sky varies during the year. In the higher latitudes, like where I reside in Montana, we are much more aware of the movement of the Sun's path during the course of a year since it varies from the very short days of winter to the very long and delightful days of summer.

If you make an observation of the rising and setting of the various planets when they are visible in the night sky, you will see that they roughly follow the same plane across the sky that the Sun does. This path is called the *"ecliptic."* The ecliptic represents the disc that the orbits of all the planets take around the Sun. If you've ever seen a picture of the Solar System you may have noted that the planets roughly orbit the Sun along the same plane. This image is usually presented with the orbits being parallel to the bottom of the page you are looking at.

So, out we go into our Montana backyard on a clear night when we can see the giant stripe of stars across the sky, which we call the Milky Way. The Milky Way Galaxy is what is known as a spiral nebula. It contains over 200 billion stars of which our sun is one. Our galaxy is shaped like a large disc of stars rotating at great speed. It has a luminous center where millions of stars are concentrated. It is estimated to be approximately 100,000-light years in diameter and

The Path of the Earth through Space.

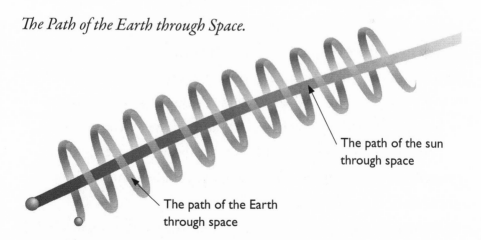

The path of the sun through space

The path of the Earth through space

16,000-light years thick at its center tapering to 3,000-light years in thickness at its outer edge. We have not yet been able to send a spacecraft outside of our galaxy to look back at it.

So, what we know about the shape of our galaxy has been calculated from what we can see of other spiral nebulas elsewhere in outer space. Just as the path the Sun takes across the sky shows us the disc of planetary orbits that make up the Solar System, so that stripe of stars across the sky, which we call the Milky Way, shows us the disc of stars that form our galaxy. When we look at the Milky Way in the night sky, we are actually looking at the thinnest dimension of our galaxy as if we were outside of it viewing it on its edge. We see the thickness of the galaxy from the position of our Solar System, which is approximately two-thirds of the radius from galactic center to its outer edge. Once we have observed both the disc of our solar system and the disc of the galaxy, we will see that the two discs are just a little short of perpendicular in orientation to each other.

As our solar system proceeds along its orbital path, the line of our orbit corresponds to the axis of rotation of the planets around the Sun. In space there is no up or down, only relationship. If we viewed the plane of our galaxy parallel to the surface we are standing on we would see that our solar system is flying through space more or less on its edge. Since the Earth is rotating around the Sun at the same time, the actual path that our planet takes through space looks like a corkscrew. To make matters a little more complex, the path of our elliptical orbit toward galactic center is curved.

The nature of this oscillation along a curved path through galactic space, plus the knowledge that the Field emanates from galactic center in a spiral form, means that our experience of the increasing power of the Field will not necessarily be smooth. Humanity is enjoying an overall acceleration of increasing inventiveness and knowledge as the Earth moves toward galactic center. But, as the ascending cycle of the ages progresses, there will also be periods of retrograde with respect to the development of the human intellect.

However, that's not the whole story. There is a second and third layer to the explanation of why our experience here on Earth is not a smooth progression into the higher ages. The Sun and some of the planets in our solar system emit magnetic fields of varying strengths. Magnetic fields result from the presence of electrical activity. The Sun's magnetic field, called the *"heliosphere,"* spirals out into the Solar System engulfing all of the planets.

The Sun and the planets also emit *auric fields.* Just as a magnetic field is the result of electrical activity, an auric field results from the presence of consciousness. The causal and astral bodies of a human being that precede the formation of a physical body are auric fields.

In Chapter 7, we discussed the concept of Gaia, the consciousness that keeps the Earth habitable for all the life forms that live here. If the concept of the Earth harboring a sort of consciousness is new to you, then I am about to confound the situation even further by claiming that the Sun and all the planets in our solar system are also home to various sorts of consciousness. Indeed, according to the ancient principles of astrology, each planet exhibits an archetypical personality that exerts its influence on all of us.

Each of the planets in our solar system has its own orbital speed around the Sun. Each planet emits various types of fields that overlap the fields of the other planets and are constantly moving. At any given moment the exact configuration of all the different fields in the Solar System is bathing the Earth with influences that are constantly changing. The result of all this movement is a kaleidoscope of overlapping influences all producing constant change within the Earth, and also more importantly, within us. This kaleidoscope of movement is the second reason that our experience here on Earth has a lot of variety in it.

In addition, there is a third component to be considered here: The Field of All Possibilities can manifest our most inspired dreams. The Field can also manifest nightmarish reality from our deepest levels of ignorance. In Part V, we'll talk about how that ignorance is responsible for the quality of the experience we are all sharing. To sum up our theory of the Creative Principle so far, the Field of All Possibilities emanates from galactic center as an etheric wind of subatomic waves of pure potentiality. Our solar system, with all its overlapping and constantly changing auric and magnetic fields is flying through this etheric wind on its edge in an elliptical path at great speed. As the Sun approaches its binary companion star, our solar system's velocity through space accelerates. The etheric wind of the Field varies in its intensity with our proximity to galactic center. The closer we get, the more powerful are the forces of manifestation.

Though we are accelerating toward galactic center, the slightly tilted corkscrew path that Earth takes through space, combined with the constantly changing, overlapping auric and magnetic fields of the planets in our solar system, plus the spiral shape of the Field of All Possibilities itself results in a multi-dimensional ebb and flow in the development of human consciousness. That ebb and flow is reflected in the affairs of humanity.

First one nation achieves superpower status then very quickly goes into decline. Then another nation ascends to superpower status only to go into decline shortly thereafter. The Romans, the British, the French, the Germans, the Russians, the Japanese, and the United States have all gone through this process. It appears that there is such a thing as a *cultural biorhythm.*

The Japanese have illustrated this concept in a very short period of time. Right after World War Two Japan was utterly destroyed. I remember how, as a child, the phrase "Made in Japan" was considered to be a joke. Very quickly, the Japanese realized that they needed to regenerate their manufacturing capability. Having lost great numbers of their most creative citizens during the war, they decided to copy the techniques of the West as the shortest path back to prosperity. They applied themselves diligently and made advances at a rapid rate. Before long the words "Made in Japan," were cropping up on goods all over the United States and the rest of the world.

Meanwhile, the United States was busy fending off the Russian threat and going through the massive cultural upheaval of the '70s. At the same time, big business was deciding that the American work force was too expensive to continue doing all the dirty work. We began to export the incredible manufacturing capability that saved us during World War Two to the so-called "third world" nations. Little did we know how the arrogance of success would come back to cripple us in the 21st century.

In less than 30 years, Japan rose to manufacturing superiority in electronics, household goods, and automobiles. Then they made the same mistake we had. They began to exhibit the arrogant attitude that often seems to come with the massive success of any endeavor. In an orgy of self-confidence they over extended their financial capabilities. By the early '90s, the world-recognized economic miracle that was Japan began sliding into decline. Japan had gone through a complete example of a cultural biorhythm.

The Japanese are a very old culture. They have a lot in their history to be proud of. One of the aspects of their national character that led to World War Two was a deep-seated belief in their own cultural superiority. Their most recent biorhythm arose out of the ashes of the war. Having been utterly defeated, they started over and created a new belief system about who they could become. They dedicated themselves to achieving that vision and went through their own version of the American Industrial Revolution in a very short period of time. Their cultural biorhythm peaked as they were recognized as a world leader in manufacturing and began to decline shortly thereafter. That decline was triggered by a re-emergence of the old belief system that they really were a superior culture as evidenced by their rapid rise into prosperity.

Starting over 200 years ago, the United States has now gone through a similar cultural biorhythm, arriving at a similar place at roughly the same time as the Japanese, and for many of the same reasons. The European countries have been taking turns shuffling through alternating cultural biorhythms for centuries. The curious thing about this moment in history is that so many cultural biorhythms are coming into decline at the same time. This phenomenon is indicating to us that something on both a global and a cosmic scale is happening.

There seems to be a common occurrence that precedes the decline of a cultural biorhythm. As a cultural biorhythm peaks a subtle shift in consciousness occurs. The cultural mindset progresses from "striving to achieve," to "have achieved." A sort of cultural ego emerges. The purpose of that cultural ego is to distinguish itself from others by declaring itself to be separate from, and superior to the rest of humanity. In later chapters we will see how this belief system of exclusivity is at the root of many of our worldwide problems.

As you may suspect from what you have read thus far, a quickening in human affairs is at hand. One of the secrets of the Creative Principle is that *what we declare to be true in our hearts and in our minds as co-creators of the universe is ultimately manifested.* This is true on an individual basis, on a regional basis, on a national basis, and on a global basis. Today there is a common perception among the people of the world that time is accelerating, and that something is about to happen. Something is not just about to happen it's already happening.

The good news is that everything you emote will manifest. The bad news is that everything you emote will manifest. Simply put, that is why we are where we are today.

Chapter 16

Cultural Biorhythms

Consensus reality is a general agreement among a group of people as to what is true about their existence. Comprised of a series of decisions about how life "is," it's a belief system held in common. While consensus reality can be shared by only two people or by very small groups (as it might be in a cult, for example), it is the larger cultural version that is of interest to our inquiry. Consensus reality varies widely from country to country. What people believe in the dictatorship of North Korea is very different from what people believe in the democracy of the United States. A consensus reality is not necessarily what is true, but what a large number of people hold to be true.

A national consensus reality is made up of the political, economic, religious, and scientific beliefs of a country, to name just a few of the types of beliefs that it may encompass. On a national level, one of a consensus reality's main ingredients is what the nation believes it should be afraid of, and take steps to prevent from happening. Most of the larger nations in the world have built sophisticated weaponry and trained large armies to ensure that other countries cannot force an involuntary change in their version of consensus reality. There's nothing wrong with honest self-defense. Aggression is another matter entirely.

In Chapter 15, I presented the idea that many nations have gone through a similar process of establishing a consensus reality and acting out the underlying belief system until it no longer serves its creators and collapses under its own weight. This phenomenon, which we see throughout history, could be thought of as a cultural biorhythm. The rise and fall of the Japanese Empire during the 20th century is an example of a cultural biorhythm. It took less than 50 years for the Japanese paradigm to run its course.

Similarly, it wasn't that long ago that the statement, "The sun never sets on the British Empire," was a fact of life. At the height of its power, the British Empire was the largest empire in history, exceeding even Rome; today it is one of the smallest. The rise and fall of the British Empire during the 19th century also illustrates the concept of a cultural biorhythm.

Think of a cultural biorhythm as a wave structure of very large amplitude. It could be a wavelength measured in hundreds of years. The wave is generated by the adoption of a consensus reality or belief system by a culture. The wave is then propagated by the projection of that belief system onto the world stage. In the case of the British Empire of the 19th century, the wave was generated by a belief in the moral and genetic superiority of the ruling class, and projected around the world by the tactical superiority of the British Royal Navy. Doesn't this sound familiar? Nearly every culture that has ever committed aggression against its neighbors has had a ruling class at the helm that believes wholeheartedly in the superiority of a certain belief system.

Radical Muslims, for example, believe that at some point the world will be united under the moral superiority of Islamic law, which openly oppresses females and condemns "unbelievers." Despite the fact that the world has never been united under any sort of "mono-culture," many Muslims believe that it soon will be. This belief system is a classic example of a closed-loop belief system, where you either accept the original premise and you're in, or you don't—and you're out. If you don't, you might be condemned in the hereafter or even be exposed to violence in the present.

The Nazis of the Third Reich believed in the genetic superiority of the German people to the extent that they were willing to exterminate entire cultures in order to "Germanize" Europe. The Germans believed that the Third Reich would last for a thousand years. It lasted only 12 years. The United Soviet Socialist Republic (U.S.S.R.) rose on the wings of lofty ideals out of the ashes of Tsarist Russia, swiftly corrupted itself, and fell in less than a century. The United States, now in serious decline, and which her leaders and citizens are nonetheless fond of describing as the "greatest country in the history of the world," has spread its brand of financial organization around the world like a virus only to watch it begin to consume its hosts.

Each one of these regional histories is an example of a cultural biorhythm. Each of these biorhythms rose out of the ocean of belief like a wave, passed over us all, and then declined, leaving massive destruction in its wake. What do all these examples have in common? In the same way an individual manifests a subjective reality through his or her body of knowing (whether or not that knowing is aligned with truth), cultures manifest subjective realities. Cultural biorhythms result from the collective consciousness of the people who inhabit the culture. This process is the essence of the Creative Principle in action.

The Soviet experiment is a particularly telling example of this point. It is not so much what we say that manifests our reality, but what we are. The Soviets said their purpose was to "honor the equality of the worker" and "save the worker from the oppression of the ruling class." So they destroyed their corrupt, bejeweled monarchy and replaced it with a new ruling class that wore cheap suits. They then proceeded to execute over 20 million people who the leaders perceived as not wanting to fall in line with the new consensus reality. Human beings fear death. Death is the dissolution of the physical form. Those executions institutionalized fear as a daily part of Soviet life.

Despite the outward declaration by the Soviet leaders that they stood for the equality of all their citizens, the real and more powerful declaration underneath their words came through their actions, which demonstrated that they believed in the liberal use of fear to rule the masses. The self-fulfilling prophecy of the Field is that whatever you think and feel intensely over time, you will eventually become.

In the same way that the Field manifests the reality of an individual, it also manifests the reality of a nation. Fear is destructive. The natural outcome of using fear to motivate others on a national scale during the ascending cycle of the yugas is the eventual destruction of the institutions promoting that fear. The practice of fearing produces more things to fear. This is why fear based cultures crumble after reaching their zenith. The higher the age, the more quickly this occurs.

I will admit that one of the factors motivating me to write this book was the emergence of the Global Financial Crisis. It is clear to me that the waves of hundreds, if not thousands, of cultural biorhythms are lined up like so many tsunamis, with foaming crests, all proceeding toward the beach of world affairs at a high rate of speed. What we have never seen before is the impending crash of so many cultural biorhythms at the same time, which threatens to drown us all. This isn't just a financial crash, it is the crash of the fear based social, political, religious, and financial belief systems that are used to run our world.

Ultimately the purpose of the Field is to keep reminding us who we really are. We have the power to create our world any way we want it to be. The Field will remind us again and again that we are the cause of what is manifesting, until we demonstrate by our actions that we understand. It should be clear by now that what is needed is an in-depth self-examination of what we say we believe about our presence here on Planet Earth, versus what our thoughts and actions demonstrate that we really do *believe*. That self-examination will culminate in us developing a new cosmology for the 21st century, one that is sustainable, and will hopefully lead us to create a new series of healthier cultural biorhythms.

You may have found the discussion in earlier chapters, describing our position on a binary orbital path within a galaxy designed for the propagation of consciousness, difficult to digest. Whether you choose to embrace Sri Yukteswar's version of galactic structure isn't the important thing. What's important is the understanding that our planet has been provided as a laboratory for us to experiment with the Field until we realize how we create our own reality, and that we have the free will to create that reality any way we want.

The power and rapidity with which our input into the Field manifests our experience is increasing. The Field is a very large mirror. The image standing in that mirror is us looking back at ourselves and wondering,

"Who have we been, and who are we becoming?"

Chapter 17

The Quickening

Science lags behind philosophy in the adoption of new ideas. As a result, there is little widely accepted science available to back up the story I've been telling you. My story is comprised of a series of interpretations about what certain scientific discoveries actually mean. Since you can't go to the halls of science to verify whether or not the message of this book is true, you'll have to make up your own mind about reality.

One of the roles of the philosopher is to suggest how science might interpret the discoveries it makes in its quest for proof of how the universe works. Often, I am not content to wait for science to verify what I've been thinking, but prefer to draw upon that which science has proven in the past to guess where we might be going as a species. Copernicus couldn't pursue proof that the Earth revolves around the Sun without first conceiving that it did so. In order to formulate his hypothesis, Copernicus had to propose an alternate explanation for what he was seeing in the sky. It was tough for people to believe him because, after all, the sun does appear to be moving around the Earth. His idea that the Earth revolves around the sun was considered to be an outrageous theory in the 16th century and yet it has since been proven to be a fact with which nobody disagrees.

The outrageous theory I am proposing is that the experience you are having, that you call life, is not being imposed upon you by the world, but is being attracted into being and manifested into form by the qualities of your consciousness. I'm also presenting the hypothesis that the process by which consciousness manifests the physical world is accelerating, and there is an astrophysical cause for that acceleration.

In pinpointing where we are in that process, I've told a story about a 24,000-year cycle of existence within which the human experiment has evolved. I'm suggesting to you that the binary orbit of our solar system through the etheric wind of the Field emanating from galactic center is responsible for setting the stage upon which current events are being acted out.

In the current age, the time between the conception of an idea and its resulting manifestation is growing shorter. We are also seeing a corresponding quickening in the dissolution of old paradigms of social, political, religious, and economic organization that no longer serve the expansion of conscious human evolution. Many people think that the Global Financial Crisis is just a particularly bad, but temporary down-turn in the stock market, which will eventually right itself. Then it will be back to business as usual. In fact, our financial mess is a symptom of hundreds of cultural biorhythms, and especially the paradigms under which they were founded, coming to a conclusion at the same time.

This unintended convergence of consequences is a direct result of the quickening that is taking place in human affairs. Now that we have an understanding of how and why this quickening is taking place, a critical review of the collective thought processes that are generating our experience is called for.

How do we go about uncovering the old paradigms that have

manifested the troubled world we live in? We can start with what has
already manifested, and work backwards to discover the underlying
belief systems that have caused the world to be the way it is today. As
we examine some of those belief systems, it will soon be evident that
they are surprisingly few in number.

Is there a primary belief system that has manifested a world in
conflict with itself, with all the old systems coming apart at the seams,
culminating in the chaos of a global financial crisis? Yes, there is. It
is called "exclusivity." We actively promote exclusivity in many ways.
When we want to advertise something as being the very best, we say
that it is exclusive. What that really means is that very few people can
afford it. It is available to only a few in the top echelons of society. The
implication behind the whole system is that we should want to be recognized as being members of that group and we should want whatever
is considered to be exclusive. To be admitted where others are excluded is considered to be a mark of personal achievement.

Let's say that you have come to the point in life where you are
able to reward yourself for your achievements by purchasing a really
expensive sports car. Off you go to the exotic car store and you decide
on a Ferrari 599 GTB. You plunk down $300,000 dollars and take the
car home. You may have bought the car because you are an enthusiast.
You revel in the smell of the leather, the feel of the seats. You become
joyful as you experience the unparalleled acceleration and listen to the
carefully tuned exhaust note burbling behind you. You are thrilled as
you feel how it sticks to the road when accelerating through a corner.
That's one way to experience the car. But there may be another dimen-

sion to why you bought it.

As you pull up to the restaurant for lunch you feel a little thrill as you observe the admiring stares of everyone who sees you driving this mark of distinction. You park the car and get out and go into the restaurant with your head held high in the knowledge that you are one of a very few individuals in the world capable of owning such a prized object. When lunch is over you return to the car and have to wade through a crowd of onlookers who want to know what it's like to own and drive such a machine. Once again you feel the thrill of being above the crowd, a man alone at the top.

On your way home, you have a rare experience: You pass another Ferrari going the opposite direction. You check out the car and its driver with interest and give a little wave. There is another man of distinction. Reluctantly you pull into your driveway and put the car away. The next day you get up, open the garage door and admire the gleaming beauty before you. You hop in and fire up the engine. It will be a glorious drive to the office this morning.

But over night something very strange has happened. You must have shifted into a parallel universe as you slept. While zipping up the on-ramp to the freeway, you notice that everybody is driving the same Ferrari you are! If in that split second your elation shifts from the glorious feeling of man and machine at its highest level of expression and crashes into a kind of depression, then you know without a doubt you believe in the paradigm of exclusivity.

Your physical enjoyment of the machine has not changed; it's the same car it was yesterday. Upon further reflection you realize that your enjoyment of the car was deeply rooted in the fact that only a few of the very best people could have this car, and that you were one of them. Now that everyone has one, your happiness is no longer exceptional, it's just commonplace. You are no longer exclusive. As you examine your motives, you realize that you actually derived your happiness from knowing that others were excluded from having the experience that only you and very few of the finer people could have. You slink back to the dealership to take the car back. It no longer serves the purpose for which it was bought.

This is a prime example of how an original paradigm quickens producing an unintended consequence. What happened seems unintended because you weren't really in touch with your inner motivations for buying the car in the first place. You realize that you did everything you did in pursuit of a feeling. You created a paradigm and pursued it with all of your heart and soul. Your personal paradigm went through the process of quickening and came to its logical conclusion. As a creator-in-training you called for exclusivity and received the manifestation of your consciousness: You were excluded from the enjoyment you were seeking. So, you woke up to the unintended consequences of your actions once again. In a manner of speaking you fooled yourself regarding your real motivations. The Field did not manifest from what was in your mind, but what was in your heart, where your body of knowing resides.

Make no mistake, I'm not against prosperity. I'm not suggesting that there is something intrinsically wrong with owning a nice car. What I am saying is that we need to examine the real motives that lie behind our actions. Sometimes those motives are so much a part of our way of being we have difficulty seeing the forest through the trees.

I've spent a lot of time sitting on the corral fence observing our horses. We have a gelding and a mare that have been together for ten years. Part of the year they graze in the pasture, and part of the year they are fed hay twice a day. When they are on the pasture they are surrounded by food and all is well. There is more than enough for each of them. When they are penned up and we throw some hay over the rail the matter is changed completely.

The pen has nothing growing in it, it's just bare dirt. So when we throw some hay over the fence they can both clearly see that there is a limited amount of feed available. The gelding is bigger than the mare, and if he can he will exclude her from being able to eat as much as she wants until he's satisfied. Every experienced horseman knows that if you have two horses, you set out three piles of hay, then watch them dance. The gelding will eat a little and then shoo the mare off the next pile and eat there for a while, then shoo her off the next pile. This form of exclusion is a matter of instinct based upon survival of the fittest. The gelding is capable of getting what he wants and does so without regard for the health and safety of his companion.

Humanity isn't so different. We've taken an instinctual fear from our primitive past and molded it into modern economic, social, political, and religious institutions. The subjective truth we live by is "There is not enough for all, therefore someone must go without." This paradigm can be seen on the evening news. When there is a natural disaster, people who are feeling anxiety that they might cease to exist tend to hoard food, water, and other supplies without a great deal of regard for the needs of others. Sometimes great numbers of people run around looting and stealing as if all laws have been temporarily suspended. During natural disasters we also see fewer examples of people who have overcome their fears and commit acts of compassion and heroism that are an inspiration to us all.

We have translated our most basic fears into economic systems that reward those who are capable of taking what they want without regard for the rest. We have translated those instinctual fears into social and political hierarchies that serve the elite before everyone else. We have created one religious belief system after another in which only certain people are saved. The world wide situation we find ourselves in has resulted from the quickening of these paradigms. The quickening is acting as a mirror, showing us who we have been. If we don't like the result, it is up to us to use our free will to change it.

Since everything in the physical universe has been created directly from consciousness, and since consciousness is infinite, then the possibility exists that there can be enough for all seven billion of us. We can give up our more base instincts in order to manifest a higher form of human consciousness. Before that can happen we need to adopt belief systems within which it is possible. We need to realize that we are all connected by the Field at the most fundamental level. The quickening is presenting us with an opportunity to see what we as individuals do to others, we are eventually doing to ourselves, and the time frame during which this equal and opposite reaction occurs is growing shorter.

We have seen financial crises before, but never a global financial crisis. We have experienced hurricanes before, but never one that

caused as much economic destruction as Katrina. We have seen tsunamis before, but never one that wiped out several nuclear reactors.
We've had tornados before but never one after another after yet another destroying town after town all across the country in a single season.
Are these events coincidental or are they a planetary immune system
reaction providing a commentary on the state of human activity and
ultimately the state of human consciousness? Is this the moment in
human history when we are faced with the choice to evolve or perish?
Will this be the time when we are faced with the *necessity* to make a
quantum leap in our own evolution? Will we realize that this is the
moment in history when the internal choices that we make will rapidly go through the quickening process and affect the whole planet in
new and unusual ways?

It is urgent for us to realize every human is engaged in the same
process of creating from consciousness. We think and we feel and we
experience the manifestation of that creative activity. Every experience we're having on our planet is the result of this process in operation. Every experience we're having individually and collectively is
the result of what we've done with our inner awareness. Despite the
different ways we go about the process, in a larger sense, we're all here
doing the same things for the same reasons.

We keep trying to prove that we're all different, and that according
to a variety of subjective interpretations, only the most deserving of
us should have the right to survive and prosper. We've built modern
world wide culture on this flawed paradigm. If we aren't careful we
may be successful in manifesting the consequence of that belief sys-

tem: massive loss of human life.

The prospect of the loss of human life is daunting, to be sure. You may have difficulty embracing the idea that the most effective thing we can do to prevent it is mostly internal. I'm not looking to create a following of true believers in the Vedic system of the Yugas, though I find the logic behind it works well for me. I provided this information to show that the power of manifestation, the rate at which your inner knowing manifests your reality, is not only increasing but accelerating.

The quickening occurring in human consciousness will aid us in making the internal changes required to minimize, and possibly even eliminate, a winnowing of the human population. The quickening is the result of cosmic forces in play and those cosmic forces reflect the intent of creation. So here we are at that point in our binary orbit that opens up whole new vistas in the development of human consciousness. What inner and outer changes will we have to make, and what else will we have to do to take the next quantum leap in our own evolution as a species?

Part IV

Spellcast

Chapter 18

The Spell Is Cast by Paper Money

In 1989, I quit the development business and moved away from civilization into the Sierra Mountains. I was disillusioned with city life. I distrusted those who implied they had the rat race all figured out, but demonstrated by their actions that they hadn't. I just didn't care about one more chapter in the relentless pursuit of a bigger house and a more expensive car. I wanted to see clearly and understand how life really worked.

I knew intuitively that the principles of creation were present in nature. I thought the closer I was to nature, the better chance I had of seeing what those principles might be. I spent more of my time in the mountains, being closer to nature than ever before. That experience made me a lot calmer than I used to be, and gradually the principles of creation began to emerge.

When I had occasion to go back to the city I was struck dumb by the contrast. City people seemed to live mostly in their minds. I felt that they were all running around like rats in a maze with no real overview of the big picture, and no desire to acquire one. Everybody was in a hurry, desperate to get to the next destination. They lived under the spell of popular culture. Their lives appeared to hover around whatever was on the television, on the radio, and in the newspaper. In the city there was no horizon to watch as the sun came up or went down. Inner peace wasn't really a topic of conversation. Nobody seemed to be aware that there was something missing. There was a lot missing for me.

When I went back home to the mountains I realized how much I had changed. I had developed an addiction to the real tactile experience of nature. To see the big view and the stars at night. To be in the woods and feel the conscious presence of the trees. To stand in the rain and breathe in the moisture. To ski into the house when the snow got too deep. I needed it. I wanted it because it gave me a glimpse of what living in a state of real inner peace must be like.

I've now spent a couple of decades living outside the mainstream influence of popular culture. I've managed to form an opinion as to why, when we enjoy fabulous technological progress in all areas of endeavor, the world is still such a mess. What is the nature of the barrier we have to break through to establish ourselves as a successful species that is not in danger of eradicating itself?

My opinion is that a spell has been cast upon us. It's a spell that is so big in scope that you can't see it unless you get up and walk out. You'll need to stay out there on the periphery for a while before you start to see the contrast. I debated long and hard about whether or not to include the information in Part IV in a book about human potential. In the end, I decided it would be impossible not to include a detailed description of the spell we are under if we intend to lift that spell and get on with the process of transforming our species.

The experts will correctly tell you the information I have provided in Part IV is an over-simplification. However, every story that is complex in its details can be broken down into simple statements about the belief systems of the participants. It can be painful to wake up and find out that the story we've believed for hundreds of years about how the world actually works is really a spell that has been cast upon us in a most shocking manner. It can be painful to wake up and discover that we have been enslaved by a certain class of people, without our knowledge or consent, because they persuaded us to hold certain beliefs. The tale of how that enslavement occurred begins with the story of paper money. In this world of duality there are two sides to every story. Every story has a positive side and a negative side. According to the Law of Unintended Consequences, paper money has a darker side, which is the subject of this tale.

Strangely enough, paper money came into common use in Europe right around the time when the ascending Kali Yuga gave way to the more highly developed age of Dwapara Yuga. Paper money had the potential to become an evolutionary invention that held great promise for the expansion of trade. One could easily carry large denominations of paper money undetected, whereas movements of large amounts of gold or silver were noticed by anyone present through the efforts exerted by its owners to move it about. In addition to that, gold and silver coins weren't always available in denominations certified as to value by local government, and the purity and weight of it varied from kingdom to kingdom. In order to ascertain the exchange value of gold and silver coinage it was necessary to determine both its purity and its weight. Paper money in small denominations made commerce so much more convenient. Here's a story about how it came into being.

Gold and silver have been in circulation since early Egyptian times. The trade of goldsmithing can trace its roots nearly to what is thought of as the dawn of human history. Then, as now, goldsmiths served the upper echelons of society by crafting beautiful golden artifacts to be worn by royalty, the clergy and the merchant class as a symbol of power. It was common to wear gold chains that could be used as money simply by cutting off a link or two in exchange for goods. The gold trade had become relatively sophisticated from the earliest times, and the goldsmiths possessed the technology both to weigh gold and determine its purity. In addition to the tools required to craft gold, the goldsmiths also needed to have a vault to store it securely, and the muscle to protect the vault from thieves.

When a merchant accumulated a quantity of gold that exceeded his daily requirements for conducting business, he needed a safe place to store it. The obvious choice was to deposit his gold with the gold-smith who already possessed the facilities needed to ensure its safety. So the merchant would bring his gold to the goldsmith and rent space in his vault. The goldsmith would take the merchant's deposit, place it in a pile on a shelf in the vault, and label it with the mer-chant's name. Then the goldsmith would make an entry in his ledger recording the transaction. The merchant could then come at any time and verify that his gold and the gold of others was still in the vault in little piles with all their names clearly visible.

When the goldsmith took a deposit in this manner he also wrote the merchant a receipt that bore the merchant's name, the amount of gold on deposit and a promise to pay the merchant his gold upon demand. When the merchant needed more money to make a pur-chase than he had in pocket change, it was still necessary to go to the goldsmith, make a physical withdrawal, take the gold to the site of the transaction, verify its weight and purity and do business. This process was cumbersome and it became clear in a very short period of time that the paper receipt issued by the goldsmith would be accepted in lieu of the physical gold by more and more people who understood that they could cash the receipt for real gold at any time.

By and by the merchants began to request that the goldsmith issued many receipts in smaller amounts than the total deposit held in the vault to make the exchange of paper promissory notes, for goods or services, more convenient to all. This the goldsmiths were very willing to do since it was clear they were beginning to make more money renting vault space to merchants than they were earning by actually making artifacts out of gold.

The more convenient they could make the use of their gold-backed receipts, the more customers would come to rent space in the vault. They were anxious to service the needs of this new form of clientele. They didn't stop being artisans, but they now understood that their businesses had two separate divisions. The receipts issued by the goldsmiths had become the first paper money. The goldsmiths had become the first bankers.

The goldsmiths belonged to trade associations known as "guilds," which would meet periodically with their colleagues from other towns and cities to discuss the secret aspects of their businesses. Merchants who were traveling were making requests to be able to deposit gold with their hometown goldsmith, carry the paper receipts to another town and withdraw gold from a colleague known to their goldsmith.

The goldsmiths realized they could charge a small fee for this service and were eager to add those fees to their profits.

As time went by it became apparent that the practice of segregating each depositor's gold into its own little pile, and labeling it with the depositor's name, as well as writing the depositor's name on his receipt was no longer necessary or desirable. The merchant no longer cared whether the gold he withdrew had been previously owned by him or not. All the merchant cared about was access to the equivalent amount of gold he had on deposit in whatever town he happened to be in. In some ways it was desirable for the merchant's receipts not to bear his name, since in that way the history of his transactions could not be traced.

The goldsmiths found that erasing identities as to the ownership of the gold in their vaults was also of advantage to them. Any depositor could come to the goldsmith's shop and still see the single pile of gold in the vault, which gave them great comfort. They could still withdraw their gold upon demand as before; it simply no longer had their name on it.

This seemed like a reasonable thing to do, making the process more convenient for everyone. As the goldsmith's business increased, and the issuance of receipts became commonplace, the goldsmiths noticed something very interesting. Now and then they could issue an anonymous receipt for gold they had not received and it would be honored at face value by all of those people who had become used to accepting paper money.

Very quickly the goldsmiths realized that they could actually create money out of nothing by writing receipts for gold they didn't possess. Nobody would ever be the wiser, since they could all come and still see the pile of gold in the vault. This was a potentially dangerous business situation because the goldsmith was issuing more promises to pay than there was gold in the vault. If a circumstance occurred, such as a war, when all the depositors got scared and wanted their gold back at the same time, the truth would come out. There was not as much gold in the vault as there were receipts to claim it. Thus, there would be a run on the bank and the goldsmith would be ruined.

To combat this problem the goldsmiths came up with an ingenious plan. Instead of writing bogus receipts to spend directly, they would create those paper receipts and loan them out to be repaid in time and with interest by the borrower. They could also require, as a condition of the loans, that the borrower put up collateral in the amount of the loan and that the debt be repaid in gold. Each bogus promissory note loaned by the goldsmith in this manner represented a promise to pay by the borrower.

If all the goldsmith's notes were to be presented for payment in gold at the same time, the goldsmith would not be able to pay, but could point to the borrowers and say they owed the missing balance. If the borrowers could not come up with the payment, the goldsmith could repossess their collateral and be made whole. These practices were, of course, fraud on a grand scale, but nobody knew it except the new banker/goldsmiths who kept the whole business secret and discussed it only amongst themselves.

The goldsmiths were deliriously happy with their newly discovered ability to acquire great wealth made from nothing. Due to the requirement that their loans be repaid in gold, the vault was bulging with the gleaming metal. The goldsmiths expressed their good fortune by adopting an opulent lifestyle. Everyone thought the goldsmith's prosperity was the result of them being good businessmen and were therefore encouraged to do business with them. But then something unforeseen happened.

After a while the goldsmith's borrowers wanted to know why they couldn't repay their loans with the paper receipts they had collected in the course of doing business. After all they bore the statement, "Will pay in gold on demand," and were touted by the goldsmith himself as being as good as gold. The goldsmiths were forced into a corner. They couldn't reveal the secret that they had been lending out worthless paper with no gold behind it or they would be ruined. They were forced to take the worthless paper back in repayment for the loans given. As the use of paper money increased in popularity, the great bonanza in loans repaid in gold was reduced back to only the interest charged on the loan.

However, the goldsmiths had become used to the newfound riches acquired by creating money out of nothing and were loathe to give up, or reduce the luxury of their lifestyle. If they reduced their opulent standard of living their depositors would take it as a sign that the business was failing and would withdraw their deposits. Greed being what it is, they had to come up with another way to continue the deception. The solution was to increase the amount of loans they issued and live off the interest income.

To accomplish this, a problem needed to be solved. Each time the goldsmith made a loan he recorded the transaction in his ledger. Each loan was recorded as a liability against the total deposit of gold in the vault. Anyone who had the authority to inspect the goldsmith's books, such as the king's tax collector, would discover that the amount of liabilities recorded greatly exceeded the value of gold in the vault, and the books didn't balance.

The solution to this problem was as easy as kissing the king's hand. All the goldsmiths had to do was to record the loans they had made as assets represented by promises to pay, instead of *liabilities* represented by the absence of gold in the vault. After all, the borrowers had promised to repay and had put up collateral in the amount of the loan. Lo and behold the books balanced again! Now the goldsmiths were free to loan as much paper money as they could find clients to borrow it and the books would stay balanced.

As the paper money loan business increased, the amount of gold in the vault represented a smaller and smaller fraction of the total amount of paper money in circulation. However, the amount of gold in the vault remained more or less constant and the goldsmith could point to it with pride and comment on how stable the economy was. The merchants noticed what a good job the goldsmith was doing, and everybody seemed to be happy once again.

Eventually the king discovered what the goldsmith was doing and summoned him to a meeting. Instead of putting a stop to these fraudulent practices the king told the goldsmith that he now had a new business partner.

After a time the paper notes began to be so common, and so easy to get, that the people began to lose confidence in them as a medium of exchange, and so they raised their prices to compensate. Fewer people came to the goldsmith's shop requesting loans and overall income declined. The people were losing faith in the goldsmith's notes. Then the king had an inspiration. He told the goldsmith to acquire the brand-new invention known as the printing press and put the king's name and his likeness on the promissory notes.

Voila, the whole business was no longer a scam! It had just become legal. The new paper money was accepted by the people with enthusiasm. After all, it bore the king's image and his direct promise to pay in gold. Therefore the people were mollified. Prices went back down, more people came to the goldsmith's shop requesting loans, and the goldsmith's and the king's income was restored. This phenomenon did not go unnoticed by the king and his goldsmith, neither of whom were dummies.

They understood that the more paper they printed and loaned, the more money they made in interest. They also understood that the process of printing more and more money, backed by a reserve of gold that only represented a *small fraction* of the total amount of money in circulation, automatically made that money worth less and less as time went by.

They knew that wealth does not come out of thin air and that the scam they were perpetuating was simply a method of draining real wealth from the people who were repaying the bogus loans and interest with the products of their blood, sweat, and tears. The inflation created from printing more money than there was gold in the vault was a method whereby the king and his goldsmith could confiscate a portion of the wealth of all the people that the people could not see—and which for the most part they didn't have the education to understand. It was a hidden tax on the people that was kept secret from them.

The king and the goldsmith discovered that the more credit they issued the more money they made and the weaker society became overall. They discovered that they had their hands on the throttle of the economy. They would push the throttle in and take enormous profits. As the people grew weak from their exertions, the king and the bankers would pull the throttle back and give them time to recover. Then they would start the process all over again and God help the kingdom if they ever went too far.

Then the king and the goldsmith noticed something else. Since they were in control of monetary policy, and could inflate or deflate the currency at will, they were in a perfect position to manipulate all the markets that used their currency and position themselves strategically so they could "buy low and sell high," making even more money.

Thus, the now commonly accepted ruse known in the financial industry as *"fractional reserve banking"* was born, and along with it the inflationary/deflationary cycle of modern business and politics.

This is the part I want you to remember. Fractional reserve banking equals a perpetual cycle of inflation and deflation. Paper money in itself is worthless. The only thing that gives it value is that it represents an agreement between people to exchange it for goods and services. But like the precious metal the paper money is supposed to represent, if you bend it back and forth too many times, metal fatigue sets in and that agreement breaks.

The Emergence of Paper Money in Europe

The pages of history are littered with stories of one great culture after another that has been destroyed by their own central bankers who promoted such schemes until the currency was destroyed. Marco Polo's tales of the widespread use of paper money during his travels to China in the 13th century could have been the influence that started the practice in Europe. Aside from the handwritten notes provided by goldsmiths, the first large scale institutional printing of European paper money occurred in Sweden in 1661.[1]

Johan Palmstruch, a Dutch merchant, is credited with the introduction of paper money in Europe. Palmstruch founded the Stockholms Banco in 1657 with King Charles X Gustav, taking half the profits. In 1661, the bank began printing and lending paper money. Palmstruch printed too many bank notes without collateral, leading to the inflation of the currency and the collapse of the bank in 1668. Palmstruch was accused of fraudulent bookkeeping and imprisoned by the king until 1670. He died a year later at age 60.

John Law was born into a family of goldsmiths in 1671. He worked his way up to hold the powerful position of Controller General of Finances to King Louis XV of France. In 1719 he told the king, "I have discovered the secret of the philosopher's stone. It is to make gold out of paper." But being a novice at this process he busily inflated the currency beyond its limits to recover, and then presided over a chaotic collapse of the French economy. John Law died a poor man in Venice in 1729.

The wise parasite allows his host to live so he can feed another day.

Since 1913 the currency of the United States has been inflated and deflated in cycle after cycle until is is now worth only a few percent of what its value was when it was first issued. Those who control the banking system have drained immense wealth from the purses of the common people. That steady loss in purchasing power has been confiscated from the people mostly without their knowledge or consent.

This financial model is now the international standard in banking systems all over the world. It is the main force behind the Global Financial Crisis, the implications of which are still unknown. You may want to ask—who are the people perpetuating this scam against society? Have they gone beyond the ability of the dollar to recover? Have they created a monster they can no longer control? In some circles they are known as the Grey Men.[2]

Chapter 19

The Rise of the Grey Men

I define the term "Grey Men" as describing those people who use money to manipulate, enslave, and in some cases physically sacrifice other people in order to maintain or increase their personal power and fortunes. To tell the story of the rise of the Grey Men it will be necessary to go back to what we think of as the beginning of time, because the Grey Men are that old. This part of our story about creation tells of the Descending Arc of the Yugas. It is followed by a short description of the Ascending Arc of the Yugas bringing us up to the present.

The Descending Satya Yuga: 11,501 to 6,701 B.C.E.

Let's visit the first age in the descending arc of our binary orbit, Satya Yuga, which was 4,800 solar years in length. If we were there now with the people alive at that time, we'd see how our solar system has rounded that point in its orbit closest to galactic center. The most recent ice age has come to an end. The giant ice caps that once extended as far as 40 degrees north latitude and 40 degrees south latitude have melted completely, and the world has been subjected to a cleansing deluge. Great Earth changes have taken place. Some land masses have risen and some have sunk. Almost all evidence of previous higher civilizations has been destroyed. What remains will be misinterpreted down through the ages.

The tales of this event have been told in our religious traditions for millennia. These tales are symbolic of the end of a 24,000-year age and the beginning of a new one.

In time the ice caps begin to form again, and the water recedes uncovering the land. The seed crop of the next class of incarnates has survived the Deluge. Though the age is descending, it is still the golden age told of in our surviving myths and legends. The myths tell us that the great mysteries of creation were understood by the human population as a matter of intuition rather than intellect. This is a pure state of being beyond what we call "thought."

During the Satya Yuga, the people have not yet evolved language or writing. Through the power of the Creative Principle they are able to converse in images. The way has been prepared. Knowledge of the great mysteries is common in daily life. Speech comes later and is only necessary once the divine telepathic abilities start to be forgotten. You may think of this period as a primitive time. It was no such thing. You may think the complexities of the lower age we now live in represent a superior way of life. I don't think that is necessarily true. The Satya Yuga was the Golden Age because life back then was natural. There was no guile in the land.

During our visit we would see that the force of gravity is reduced. Matter is in its most etheric state. The average stature of a human is at its peak. The average life span of a human is 300–400 solar years. The trees bring forth their fruits in balance for all humankind. It is a time of great harmony and joy amongst the gatherers. There is no need for money, as no medium of exchange is required to live. Life is simple and abundant. There is no fossilized evidence of these people because upon their deaths the loosely knit atomic structure of their bodies dispersed like gas. The Satya Yuga is represented in our current mythology as the story of the Garden of Eden.

During our visit we would see that humanity has not yet turned to predation. There is no need for civilization and no need for a written record. The planetary climate is a reflection of the consciousness of the people. There is no architecture from this period, as none is needed. As in the animal kingdom, no one is late and no one is in the debt of another. The story of the people and their traditions are handed down telepathically. These are the times that will be barely remembered during the lower ages to come, as times of ancient mythology when gods and the giants walked upon the planet.

The human intellect was fully developed, although the Satya Yuga was a descending age.

The Descending Treta Yuga: 6,701 to 3,101 B.C.E.

Let's continue now to visit the past with a visit to the second age in the descending arc of our binary orbit, The Treta Yuga, which was 3,600 solar years in length. Though we are in a state of deceleration, the distance of our solar system from galactic center is increasing. The Field is growing weaker. Matter is becoming denser. The average stature of a human is declining. The average life span of a human is reduced to 200 to 300 solar years. Knowledge of the great mysteries of creation is still available, as it always was, but because the ages are descending, memory of the mysteries begins to be lost to the general population. The fruit that grew freely for all in the golden age is slowly dying out, requiring us to participate in our own evolution. The planetary climate is a reflection of the consciousness of the people. There is now a need to clothe and shelter ourselves.

The fear that gives birth to all other fears arises.

There may not be enough for all and we may cease to exist, and therefore someone must go without.

The difference in consciousness between the Satya Yuga and the Treta Yuga is symbolized in history as the *fall from grace*. The *fall from grace* began when humans felt fear for the first time and it has continued to this day. It was not a matter of sin against God's laws, but a natural function of our planetary system. Creators-in-training must have a thorough understanding of what we don't want to create in order to know what we do want to create.

It was during this age that the Grey Men took their place in history. They were the leaders who decided who would have and who would not have. They developed the belief that there was not enough for all and made that belief the basis of their knowing. They arose as a class of men who set themselves above all others. They established a tradition based upon the fear that there would not be enough. They devised methods that would enable them to take a little from everyone else, so that they could be sustained at a higher standard of living than the rest of the population and would never need to fear scarcity themselves. Their descendents were the kings and goldsmiths who invented paper money and banking. They still live among us today. Their belief system is a very old one, and each generation trains the next to take its place.

Let's continue our visit. Under the leadership of the Grey Men in the Treta Yuga, the people start to organize themselves into tribes so that they may compete more effectively for the sustenance that is available. The tribes make themselves different from each other in appearance and emphasize those differences. Where there was harmony before, a sense of separateness arises in the human heart. The newly organized tribes are losing the memory of the powerful natural beings they once were.

Conflict begins to stir in the land. The pure state of being that once existed has deteriorated, and the populace must learn to exercise their creative powers in different ways in order to survive. It is the beginning of humanity's descent into the density of matter—the beginning of our lessons in the structure of the universe. Telepathic abilities are fading away. Speech is developed so that the people can continue to share information and discuss their experiences. Writing comes soon afterward, in pictures that are dim representations of the telepathic images of the earlier golden age. People no longer understand each other as they once did.

Here the people learn to use their innate powers of creativity to organize against each other. A sense of the need for self-defense begins to emerge. In time, great warrior societies appear and with them, warrior kings. The warrior kings are of two kinds. The true kings understand that their power comes from a sacred trust to guide, educate, and protect the people. They understand that if the people turn against them, they will be nothing. The Grey Men are the kings who place their opulent lifestyle above the people's right to exist. They endeavor to control, enslave, and manipulate the people for their own personal gain. They do their best to cast a spell of belief over their people to ensure that the people never turn against them. The people cannot oppose what they do not understand.

The rule of spiritual law that was once carried in the heart and in the mind must now be written, so that it will not be forgotten. But writing is still very primitive and inaccurate, so the knowledge of who the people really are begins to die out.

In the Treta Yuga, the human intellect is only three quarters developed.

The Descending Dwapara Yuga: 3,101 to 701 B.C.E.

The third age in the descending arc of our binary orbit, Dwapara Yuga, was 2,400 solar years in length. Visiting this age we can see that our solar system is still decelerating and very much further away from galactic center, the source of the Field. The universe appears to be running down, while the pull of gravity that keeps our solar system in its orbit has become more powerful.

The average stature of a human declines even further. The average life span of a human is reduced to 100 to 200 solar years. The passage of time slows and the material world becomes even denser than before.

War is considered to be inevitable and mass killing is pursued as an art form. Those who are the best at making war are heralded as the heroes of society. Many of the true kings are conquered by the Grey Men. The consciousness that there was harmony on the planet at one time has long been forgotten. Money is invented as a way to expand trade. The Grey Men learn that to control money is to control the people.

The rule of law and principle that once was designed for the betterment of humanity decays into rule by decree. The rules decreed by the warrior kings must now be enforced. The Grey Men, who are the greatest warriors, are also the holders of gold. They wear it as a symbol of their power. When a leader arrives somewhere in his chariot, resplendent in shining gold, the people throw themselves on the ground and cover their eyes. They are told that "those who fear gold are blessed." Blessed are those who do not question authority" (and cause the leader the bother of executing them).

The people believe it is natural that they should be ruled by a leadership who hold the power of life and death over them. They believe it is good to fear. Manipulation of the masses for the gratification of the few is claimed as a divine right by the Grey Men. The people agree. There is a sense of great loss among the people, although they do not know of what. Life demands more from them than they feel capable of producing.

In this age, the human intellect is only half developed.

The Descending Kali Yuga: 701 B.C.E. to 499 C.E.

The fourth stage in the descending arc of our binary orbit, the Kali Yuga, was 1,200 solar years in length. This period was commonly referred to among the people as the Dark Ages. If we visit it, we can watch as our solar system begins to approach that point in its orbit that is furthest from galactic center. The Field is approaching its weakest state. Gravity, the force that keeps Earth in its orbit is at its most intense, is trying to pull our planet back toward galactic center.

The passage of time has slowed to a crawl. The average stature of a human has declined still further. The average life span of a human is reduced to 50 to 100 solar years. It is a time of blind patriarchy, when even the benefit of written law is lost. The Grey Men rule with an iron fist. The noble societies that were formed from higher principles have been corrupted and destroyed.

One by one the great societies fall. The Sumerians, the Egyptians, the Greeks, and finally, the mightiest society of all, the Roman Empire, decay and fall apart. In any age such destruction is foretold by a debasement of the currency. The ruling Grey Men and the people have forgotten who they really are and the most evolved societies are destroyed from within in an orgy of corruption.

When the world is in chaos, the great masters appear (individuals like Jesus and Buddha, who watch over the fate of humanity). They come before the darkest hour, that humankind may have an example to carry it across the void of darkness and establish itself in the ascending cycle. They come and are heralded as the prophets of the ages.

However, humanity has been reduced to its lowest intellectual level and cannot fully absorb the ancient wisdom. Many versions of the divine knowledge are written, but it is all written in an imperfect age.

The Grey Men see and understand that the prophets are a threat to their authority and they heavily influence the stories being told by the emerging churches. They decree that they are chosen directly by God to rule the people. The age is descending into chaos. The Grey Men explain to the people that God has judged them and is not pleased. They must obey and work harder to please God.

Under the leadership of the Grey Men, each tribe claims that its own scriptures and only its own scriptures were written directly by the hand of God. The people argue and make war over which scripture is right and which prophet is the one true prophet. Chaos reigns, and disease and pestilence take hold in the land. It is the Kali Yuga, the dark age of the world. The people pray to their gods asking why they have been forsaken. The human intellect is only one quarter developed.

Finally our solar system rounds the point in its orbit furthest from galactic center. We begin to turn back toward the light. Just as winter gives way to spring, once again the strength of the Field begins to increase.

The Ascending Kali Yuga: 499 c.e. to 1699 c.e.

The ascending Kali Yuga was also 1,200 years in length. At this time, the Greek and Roman empires have perished. The land is ruled by savages. Might is right and therefore whatever someone can physically do to another is considered okay to do. The belief that there is not enough for all, and therefore someone must perish, has reached its peak of expression. Barbarians have roamed the world conquering, slaying, and enslaving the common people who were unable to resist the spell cast by the Grey Men. Slowly, the strength of the Field increases and the principles upon which the great societies were founded begin to re-emerge.

The world is traveling back toward the light of galactic center and humanity begins to create again. The peaceful exchange of goods and services becomes more and more common. Though the land is still embroiled in war, humanity begins to innovate. The great sailing ships appear. We learn that the world is round and become able to find our way around it. New lands and new peoples are discovered. Trade opens up. One invention after another improves the human condition.

Finally, the end of the Kali Yuga culminates in the Renaissance. Rapid advances in literature, art, architecture, and science affect all of the people. As the people become more and more educated, the Grey Men feel they are losing control and take steps to increase their power over the people. They still feel that there is not enough for all so someone must perish. As it has been since the fall from grace, the Grey Men believe it is their right to decide who will have and who won't. Though they try to hide what they are doing, they do their best to reinforce the spell of belief that gives them influence, which remains: "Blessed are those who fear the holders of gold."

The Ascending Dwapara Yuga: 1699 C.E. to 4099 C.E.

Our current age, the ascending Dwapara Yuga, is 2,400 years in length.

Only 313 years have now passed. The strength of the Field is increasing measurably day by day. The stature of human beings is increasing. The average life span of a human is also increasing. In the past 300 years the technological advances that have occurred have doubled, and doubled again, and then again. The great sailing ships, the airplane, the automobile, space flight, the atomic age, and the digital age have left their marks upon us. It is common for the people to declare how fast time is moving. The quickening is being felt by everyone.

Every day more people are discovering our roles as creators-in-training. We are all about to make a quantum leap in the evolution of Homo sapiens. Not everyone understands it yet, but as the power of the Field increases each of our belief systems is quickly manifesting its logical conclusion. The Grey Men have already gone underground. They prefer that their belief system remain a secret lest it become known that:

They do not believe in the success of all humanity and are willing to sacrifice others for their own gain.

Anyone can see clearly there is not enough for everyone to live the opulent life style the Grey Men have become accustomed to. They are destroying the Earth with their greed and ignorance. The question that is rising in the hearts and minds of all the people is:

Why do the Grey Men have the right to decide who will have and who will not?

Chapter 20

Modern-day Grey Men

Who are the modern-day Grey Men? In order to find out, let's "follow the money." It is said that the top 5 percent of society owns a staggering 80 percent of all the wealth on the planet, and that, of this group, a fraction of the top 1 percent of society owns 50 percent of everything. Those statistics mean that 95 percent of the people on the planet are left to scramble for the remaining 20 percent of all wealth, resulting in perpetual unrest. I suspect that if we were able to find out the whole truth, the numbers would be skewed somewhat higher in favor of the top 1 percent. The modern-day Grey Men inhabit the top 1 percent. The distinction that separates them from the rest of the wealthiest people on Earth is that the Grey Men are willing to manipulate, enslave, and even sacrifice large numbers of people for their own personal gain. They are willing to send our sons and daughters to war to achieve their ends.

The Grey Men operate like a fraternity. Bound together by a philosophy, they carefully vet and train their understudies. Their top priority is the maintenance and expansion of their fortunes and the acquisition of power over others.

The Grey Men are an ancient breed going back many thousands of years to the fall from grace outlined in the previous chapter. They consider themselves to be "world men," and indeed they are. They travel the world unimpeded by the laws of any nation. They have no loyalty to nations, governments, or people. They are loyal only to the wealth and power that flows from their activities. But power does not occur in a vacuum. In order to have power over world events, one must have power over others. It is said that the kind of power that causes whole nations to tremble is the most addictive substance on Earth and that "absolute power corrupts absolutely."

> *"The few who can understand this system will either be so interested in its profits, or so dependent upon its favors, that there will be no opposition from that class, while on the other hand, that great body of people, mentally incapable of comprehending the tremendous advantage that capital derives from the system, will bear its burden without complaint and perhaps without even suspecting that the system is inimical to their interests."*
> —Baron Amschel Mayer Rothschild

The concern of the Grey Men is not power over the self. They are not seeking spiritual evolution. Having arrived at their lofty financial perch, looking down on the rest of humanity, they believe they are an evolutionary success. They think they are the pinnacle of human achievement. They believe they and their forbearers have played the game of life correctly and see this fact clearly reflected in the power they wield and the gilded station they occupy.

The Secret Money System

In 1805, one of the Grey Men, Baron Amschel Mayer Rothschild,

said, "The few who can understand this system will either be so interested in its profits, or so dependent upon its favors, that there will
be no opposition from that class, while on the other hand, that great
body of people, mentally incapable of comprehending the tremendous
advantage that capital derives from the system, will bear its burden
without complaint and perhaps without even suspecting that the system is inimical to their interests."

In the two centuries since he said that, the Grey Men have invented one more highly complicated scheme after another to segre-
gate the wealth of the world and bring it under their control. They
have been more highly successful at their endeavor than anyone in
recorded history.

Let me assure you, the Grey Men do not work thousands of times
harder than you do, though they possess thousands and even millions
of times more money and assets. They are not millions of times smarter than you are. They haven't earned the massive amount of wealth
they enjoy. In order to understand how their financial dominance has
been accomplished we need to redefine some terms. We need to draw
a distinction between the term "creating wealth" and the term "making money."

Creating wealth means to combine resources with creativity thereby producing something of value that did not exist before. Real wealth
is accumulated as compensation for engaging in the creative process.
To create is the divine mandate of the Creative Principle.

Making money used to be synonymous in the popular lexicon
with creating wealth. But this is no longer the case. Should we define a

term according to what we think it should be, or according to what it has become? Making Money now means the literal act of printing currency and the process of confiscating wealth by institutional trickery.

The Grey Men want you to believe that Wall Street and the financial markets of the world are engaged in the process of creating wealth. They trade in companies that do actually create wealth and want you to think they are doing the same thing by association. But the world financial markets are primarily made up of highly complex legal methods of confiscating and redistributing the wealth of others. The complexity of these methods is the way they hide what they are really doing from the rest of the population. The more complex they make the financial system, the harder it is for common people to understand. This ruse is a part of the spell cast upon humanity to obscure the real nature of world finance and the intent of the Grey Men.

The Prime Directive

If you've ever watched the television show *Star Trek* for any length of time you are familiar with the Prime Directive. Captain Kirk and his fellow space travelers possessed the power of warp drive (the ability of interstellar travel), which allowed them to explore the far reaches of the Galaxy. In their travels, they encountered many civilizations that were less developed than their own. Starfleet's General Order 1, the Prime Directive, was issued to prohibit those who possessed warp drive from directly interfering with the evolution of civilizations that were more primitive lest the balance of the universe be upset and those civilizations come to harm.

In the show, the Prime Directive was the most prominent guiding principle of the Federation of Planets. Such civilizations must be allowed to pursue the cause and effect relationship of the use of their own free will.

The design of the universe includes the fact that every human being is made in the image of the Creative Principle. The intent of creation is that all humans will use our innate creative power to rediscover who and what we really are. We are all, without exception for race, creed, color, or national origin, intended to understand the Creative Principle, the entity upon which the very existence of the universe depends. That being the case, the Prime Directive of our new cosmology is:

"As much as possible, do no harm to others."

Translated into behavioral terms this means each one of us should refrain from interfering with any other person's ability to express his or her divine right to the free will to create, to observe the effects of that creation, and, thereby, to evolve toward a higher understanding of universal principle through self-knowledge. If every person honored this principle in his or her relationships with every other person, what a wonderful world we might create together.

The founding fathers of the United States had it right when they wrote the Declaration of Independence, which reads:

"We hold these truths to be self-evident, that all men are creat-

ed equal, that they are endowed by their Creator with certain unalienable rights, that among these are Life, Liberty, and the pursuit of Happiness."

Obviously this idea has been completely ignored by big business and government under the direction of the Grey Men. In comparison to the average citizen of the world, the Grey Men are a highly educated group of individuals. The average citizen of the world has no idea the Grey Men even exist, or that his or her own daily life has been negatively affected by their actions. Unfortunately, the Grey Men do not subscribe to the Prime Directive of our new cosmology. On the contrary, they revel in the fact that they continue to enslave the world population by being in control of the money supply.

The Grey Men have enslaved us all by steadily confiscating a portion of our purchasing power. As Rothschild said in 1805, "Let me issue and control a nation's money and I care not who makes its laws." Since 1913 the currency of the United States has inflated by more than 1,000 percent. The perpetual inflation of the money supply is the mechanism the Grey Men have created to confiscate an obscene portion of the world's wealth. When one confiscates the wealth of another, one is preventing that person from enjoying the exercise of his or her own free will.

In the past, slaves were physically prevented from having the free will of self-determination. When the Grey Men realized that modern society had evolved to the point where such a thing would no longer stand, they invented sneakier ways to continue to feed on the energies of the common people.

Chapter 21

The Long Con

The Grey Men always and everywhere cast the spell on the rest of the population that the universe is a top-down hierarchy beginning with a judgmental god. You are supposed to believe, therefore, that it's natural for your life to be managed by other people in authority, who are so far above you that you have no hope of ever influencing them to do what you think is the right thing to do.

If you travel to New York City or Washington, D.C., what you notice right away is how stratified society is. You can no more walk into the life of a Grey Man unbidden, than you can walk into the Vatican and demand an audience with the Pope. The Grey Men own beautiful mansions in gated communities in the world's most desirable locations. They own jet airplanes and can go where you cannot go. They isolate themselves from experiencing the squalor of the lower classes. The Grey Men are the ones riding in the limousines. When they travel from city to city, they live in the tallest and most imposing buildings. They are mysterious, untouchable, and exclusive. They have secrets, lots of secrets.

The Grey Men understand that nothing on the physical plane is more pervasive than money. Everyone uses money. It's in your bank account. It's in your pants' pocket. It's even stuck between the cushions of your sofa and under the seat in your car. You literally cannot move without using the money supply. If you want to get in your car and go someplace, you have to pay those in authority who are miles above you for the privilege of putting gas in your car. They decide the price; you don't. Except for a very few lucky folks, you have to pay those in authority for the house you live in. The price of your housing as a commodity, and the confiscatory amount of interest you must pay to obtain it, is decided upon by others, not by you. When the Grey Men exert control over the money supply, they exert control over you.

As it now stands in American life, everything we eat, wear, and use comes from somewhere else, far away, and we must use money to obtain it. This feature of modern business has evolved by design. In times past we could have been self-reliant in many more ways than today. We'd have produced our own meat, milk, and eggs. We'd have done our own baking. We'd have grown a garden and canned provisions for the coming winter. We would likely have sewn our own clothing; or someone close to us would have.

Now, the vast majority of us are self-reliant in only one way. We live to earn money so we can buy what we want and need, which is produced by others living in some far-off location. In the United States, we've lost the awareness that we can and should manufacture everything we need to live comfortably.

The modern lifestyle enslaves us to the requirements of the money supply. If money is easy to get, we prosper. If money is harder to get, we suffer. No matter what we use money for, there is someone behind the curtain to take a cut of everything we do. The story of how this has been accomplished is so complicated that, as Baron Rothschild admitted, the average person is completely unaware that the system exists, or that it has been imposed upon them.

The Long Con

The Long Con is one of the most insidious inventions ever created by humanity. It has been run in country after country for hundreds of years. It is so long that the right to run it is passed down from one generation to the next. Because it is so long, we are born into its frame of reference, much like the fish that lives in water and doesn't know it. When asked how the water is, he replies, "Water? What's water?"

The Long Con is a little different every time it is set up. It varies according to the climate, demographics, primary religion, ethnicity, and available resources in the location in which it is being run. While its aspects differ according to location, the paradigm that runs it is always the same:

There is not enough for all to live the lifestyle that I have become accustomed to, and therefore you will go without.

As it is operated today, this value system is simply a more sophisticated version of the paradigm that was used by the barbarians during the dark age of the Kali Yuga: *Might is right* and therefore whatever one can physically do to another is considered to be okay to do. The only difference is that today the *might* tends to be more financial than physical, which makes it even more insidious. If someone is coming at you with a weapon at least you have an idea of what needs to be dealt with. The system the Grey Men use is so subtle and complicated, and varies so much from location to location, that it can be very difficult to understand.

I will try to explain the Long Con here in an allegorical way.

The Long Con Phase 1: Establish the Bank

While the king was necessarily confined to the geographic location of his kingdom, the goldsmiths discovered that they were not. They were free to duplicate their fractional reserve banking scheme anywhere they traveled. Since different kings were always waging war against one another, there was no shortage of underfunded monarchs in the land. The goldsmiths would target their next kingly conquest in any country that had just finished fighting a costly war and/or where the monarchy had recently changed hands.

The goldsmiths knew that the king couldn't ever seem to collect enough taxes to keep his government in the style to which it had become accustomed. The easiest way to endear themselves to the new king was to show him that if he went into partnership with the goldsmiths, in time he could have all the money he needed or wanted. For his part, the king could see that the goldsmiths were very rich and so he was eager to follow their instructions and learn their strategies for amassing money.

The first thing the goldsmiths required from the king was to be allowed to open a central bank. The central bank is a private enterprise that has the right and ability to print the king's currency and regulate the supply of paper money. The king was very pleased to hear that the goldsmiths would pay for this entire operation and that it would cost him nothing to set it up. So, in a demonstration of power and wealth, the goldsmiths built an imposing new building with a great vault that all could see as a symbol of the king's wealth and power. They brought in an impressive amount of gold and put it on display inside the vault. Then they set up their printing press in the basement and got to work printing the new paper money with the king's likeness on it.

The goldsmiths then went around to all the merchants and told them they would store their gold in the vault for a very small fee. Any time a merchant wanted, he could exchange his gold for the king's paper money, which was a legal tender with which to transact business. This new way of doing business was beginning to catch on everywhere and almost no one wanted to deal directly with the complexities of handling gold any longer. Paper money was so much more convenient to spend. The goldsmiths told the merchants, "Not only will we store your gold for you and exchange it for the king's paper money, but if you ever want your gold back we will pay it back in exchange for the paper money you now hold, upon demand. By the way, if you need a loan to expand your business you can also borrow the king's paper money at a very reasonable interest rate."

As the vault began to fill up with the merchant's gold, the goldsmiths very quietly transferred the gold they had originally "seeded" the bank with to a new vault in another location for safekeeping. The existing vault was filling with merchant gold and the merchants were coming in to borrow the king's paper money. Business was booming. The king and the goldsmiths were happy. Phase 1 of the Long Con was complete.

The Long Con Phase 2: Print and Lend Out the Paper Money

The goldsmiths knew that paper money was starting to be the new standard of exchange around the world because it made business so much easier to conduct. As they carefully watched the increasing supply of gold in the vault they began to see two things. First, after a period of time the amount of gold in the vault would stabilize at a certain point unique to each kingdom, and second, even though some gold was being deposited and some gold was being withdrawn, the amount of gold in the vault never varied by more than 10 percent of the total. This was a critical observation since it meant that the goldsmiths could lend ten times the amount of paper money than there was gold in the vault and still be able to service their promise to pay gold upon demand in exchange for the king's paper money.

Another way to say the same thing is to say that the goldsmiths now understood that at any one time they only needed to maintain a fractional reserve, in gold: just 10 percent of the total amount of loans issued.

It is important to understand that when the goldsmiths printed paper money in the basement, even though they said it was backed by gold, that paper was essentially worthless. It only acquired value when it was lent out to a borrower because, upon signing the loan documents, the paper money was now backed by the blood, sweat, and tears of the borrower's labor plus whatever collateral the borrower was putting up to secure the loan.

Now the goldsmiths eagerly set about the business of turning the total amount of their gold deposits into ten times that amount in loans. Business was booming. The new loans were helping the merchants to expand their businesses and reach out beyond their historic limits to engage in a wider and wider sphere of commerce. The principle and interest on all the loans created out of nothing was pouring in.

As the goldsmiths and the king watched their money business grow, they noticed something else very interesting. When the merchants came in to apply for their loans, they were required to tell the goldsmith/bankers what they wanted the money for. These merchants were the movers and shakers of commerce and would willingly give the king and the goldsmiths advance notice of every big business deal they were about to engage in.

This was wonderful news, both for the goldsmiths and the king, because they could use this information to position themselves to take advantage of one new economic opportunity after another. The king and the goldsmiths capitalized on this information by using their ill-gotten gains to branch out into one legitimate business after another with which to increase their fortunes. Their position in legitimate business helped to obscure the origin of the fortunes they enjoyed. So far this information was kept secret by the king and his partners, but there was a problem looming on the horizon: Their business interests were growing so fast, and were so huge in scope, that both the king and the goldsmiths needed a lot of help just to keep it all moving. It was beginning to be more and more difficult to keep the business secret. As a result, the kings and the goldsmiths were forced to take some of their employees into their confidence.

The Long Con Phase 3: Buy the Government

The amount of business being transacted was way more than the king could handle himself. As the kingdom grew and grew, so did the layers of bureaucracy needed to keep it running. The kingdom was growing a government. The king could see that the top members of his government who handled all his affairs were beginning to understand how the secret system worked.

The king realized that in order to maintain the secrecy that was so critical to the success of his ventures, he needed to bring the key players within his inner circle. The method for doing that was very simple. The king would send for each official in turn and tell them, "You have been doing a great job and I am going to give you a bonus for the excellent work you've been doing. In addition to that, I am going to tell you one of my secrets. If you invest your bonus in a certain business venture that my friend the goldsmith is starting, you will triple your money right away."

The officials caught on very quickly and were easily persuaded to keep the king's secrets and share in the spoils. They could see that their personal fortunes existed only with the king's approval and were anxious to please him. Phase 3 was complete. The loyalty of the government had been purchased and any desire to leak the king's secrets was stopped cold. Every new member of government was educated on how rewards could be obtained.

The Long Con Phase 4: Inflate, then Deflate, the System

Over time the king and the goldsmiths noticed that when they increased the number of loans they were making with fake money, the amount of real money they were confiscating from the system in principle and interest would increase. The distinction being that fake money is money created out of nothing and real money is money that represents the toil and hardship of the borrower.

They also noticed that as the money supply inflated there were more and more dollars in circulation and each dollar was therefore worth less when traded on the open market. This fact was not lost on the merchants, or the people, who saw very clearly that they had to do more work this month to pay for the same thing they bought with less work last month—although they didn't understand why.

The king and the goldsmiths were intelligent people and they began to see they could only push the system so far before the people became so tired and so disgruntled that they would take up arms against them. The stories of these types of insurrections were being passed from kingdom to kingdom as a warning, and were considered with great interest. The king and the goldsmiths found that once the system reached its maximum level of inflation they could deflate the currency by restricting the amount of new loans that were available and they would still make money.

Once the population had come to rely on the king's credit to do business, if that credit were withdrawn, businesses would fail and jobs would be lost. Both the king and the goldsmiths found that they could make money during the deflationary cycle by repossessing and buying up the assets of failed businesses for pennies on the dollar. The king would tell the merchants, "The economy is bad, there is nothing we can do about it. But we will fight the good fight on your behalf."

When the deflationary cycle had gone on for long enough the credit window would open up and the king and the goldsmith would start the cycle all over again. The king would then rehire the people who used to own his newly acquired businesses and who were grateful to have a job. The merchants would be relieved that the recession was over, and in their zeal to do business once again, would soon forget about their recent troubles.

The Long Con Phase 5: The Switch and the Payoff

Surprisingly, the goldsmiths hadn't been entirely honest with the king. Unbeknownst to King A, the goldsmiths had been running the secret system in partnership with Kings B and C, and had never mentioned that fact to any of the other kings. Over the history of their activities, the goldsmiths had noticed that the cycle of inflation and deflation could only be run so many times before faith in the king and his currency failed—at which point the paper money system would self-destruct, taking the king and his culture with it. The goldsmiths were smarter than the king, so they prepared for this event. They were ready to profit from political and economic instability.

At the moment there was peace in the land. However, as in the past, the people were growing tired from constantly being fed upon and discontent was brewing. The goldsmiths then made a secret investment in the tools of war. They engaged a cadre of agents to go around buying up all the companies that sold horses, arrows, swords, and armor. These businesses were available at relatively cheap prices since it was a time of peace. Then the goldsmiths went to King A and told him they had friends in high places within kingdoms B and C and so had heard that those kingdoms were plotting the overthrow of King A.

The goldsmiths then sent their agents to King B and King C, and told them that King A was plotting against them. The goldsmiths had proven their loyalty to all the kings by warning them that these events were brewing. But the kings were in luck. Since the goldsmiths had foreseen these dastardly deeds unfolding they had made a sweetheart deal on all the munitions each king would need to fight his war. The goldsmiths told each king in turn that friendship was one thing, and business was another. They would be happy to sell each king all the munitions he would need to fight his war with one stipulation—since we both know that our paper money is worth nothing, dear king, you'll have to pay in gold.

This was fine with the kings who had amassed large quantities of gold during their partnership with the goldsmiths. After all, they were being attacked from outside their borders and if they couldn't defend themselves there would be no kingdom and no gold left anyway. So the kings bought all the munitions they needed in exchange for their gold.

The goldsmiths collected all the gold from all three kingdoms and rapidly shipped it to vaults in kingdoms D, E, and F where the cycle of the Long Con was only beginning. The goldsmiths told each king that they needed to close the banks and remove the gold belonging to all the merchants for safekeeping against the potential ravages of war. The goldsmiths retained title to their bank buildings and promised each king to return after the war to help celebrate their great victory and to re-establish business as usual.

When tensions had mounted to a high level, the goldsmiths told their agents to put on the uniforms of the weakest king, then go burn down the granaries of the other two kings and to make sure that they were seen.

Meanwhile, the people were starting to grow restless. They had smelled a rat for a very long time. In anticipation of the coming war, prices were starting to escalate dramatically. The king countered this development by instituting wage and price controls in an attempt to stave off total chaos. The people, finding it difficult to survive under those conditions, began to cheat on their taxes and to deal in a black market for what they wanted. This was putting a crimp in the king's finances. Finally the king instituted martial law and placed his soldiers in control of the everyday activities of the people.

When the granaries were burned to the ground each of the kings summoned his people. They each gave a great speech in which they told the people that the "bad guys" were coming and asked them, "Who will stand with me and fight to protect our land and our loved ones?" Everyone was possessed by patriotic fervor and volunteered to fight and die for king and country. The kings then outfitted the people with their newly purchased armor and weaponry and sent them out to fight and die for the homeland, as they always do.

The goldsmiths had taken enormous profits from the sale of all the war material they had paid for with freshly printed paper currency bearing the king's likeness. They had swapped the war material for the king's gold in all three kingdoms and they had removed the gold on deposit in the banks for "safekeeping."

They didn't really care who the victor would be, since they would come back to the victorious kingdom and set up shop again with the blessing of the winning king. They would also go to the defeated kingdoms and set up shop again under the protection of the winner. The Grey Men then took a short and highly self-indulgent vacation at their palace in a neutral territory and then went to work in kingdoms D, E, and F while the war was being fought. When the infrastructures of kingdoms A, B, and C were destroyed, and the winning king had been determined, the Grey Men returned to kingdom's A, B, and C to rebuild since they just happened to own the only construction companies big enough to do the job.

That, dear reader, is how the Long Con has been played by the Grey Men in country after country since the inception of paper money. Fractional reserve banking is the device they have used to confiscate the world's wealth and place it under their control. Fractional reserve banking under a central bank system is a Ponzi scheme and a doomsday mechanism. The Long Con that relies on it can be played in an infinite number of variations in any country, but sooner or later, after repeated cycles of inflation and deflation, the currency is destroyed, the culture is destroyed, and the common people, as always, pay the bill with the sweat of their brows and the lives of their young people.

I didn't want to believe that such people existed in this day and age, but indeed they do exist. Make no mistake about it: The Grey Men know exactly what they are doing. The ones at the top understand they are committing genocide and creating massive suffering so they alone can enjoy lives of excess and prestige.

In the minds of these men, the common people are expendable, as they've "always been." Finally, fractional reserve banking is now the basis of the worldwide financial system and the reason the current financial crisis is global in scope. Before we consider what the the end game of all this activity might be, let's take a look at how the Long Con has played out in American history.

Chapter 22

The Long Con in American History

The Long Con can clearly be seen within the annals of American history. The Founding Fathers were well aware of it, and many of them left warnings behind regarding the dangers of fractional reserve money and a central banking system. My purpose in writing this section of the book is not as much to expound on the financial history of the United States as to create an awareness of how the spell of the Grey Men has been cast in American history, and to reveal which phase of the Long Con we are in now. Then we can begin to talk about solutions.

If you have any doubts whether the story of the Long Con provides an accurate description of what has happened in the United States, then perhaps this commentary on the activities of the Grey Men in our nation will bring those doubts to an end.

The early colonists didn't bring much money to the New World since there was no place to spend it. The first business activity that generated wealth in the British Colonies was the growing of tobacco to be traded with the Indians and exported back to Britain in trade for supplies. In 1690, the Massachusetts Bay Colony issued the first paper money to appear on the North American continent.

As we floundered about seeking a solid political and economic footing, early banks came and went, paper currencies were issued by the individual colonies, and these currencies frequently failed. There were many boom and bust cycles. The Long Con had been in operation in Europe since 1661, and the European bankers took notice as the new American Colonies started to develop some economic activity. When the time was right they would bring the Long Con to America.

The Revolutionary War was fought to rid the colonies of the financial tyranny imposed upon them by the British Empire. Europe was already locked in the embrace of the Grey Men. A great deal of suspicion arose around any attempt to set up a financial system in America that was backed by Europeans. While the United States had just been formed, each of the member states still possessed a strong individual identity with regionally based opinions on how the whole conglomeration should be run. The government of each state preferred to run its own banking system.

However, the colonists soon made a potentially fatal mistake: They embraced the European practice of issuing fractional reserve currencies. While a viable central bank was still far in the future, the Long Con had quietly established itself on the shores of the new nation and only a few people knew it. Each state might have several kinds of bank notes in circulation. The relative values of these currencies were affected by the perceived strength of each issuing bank and the state of the local economy. This created numerous problems trying to calculate exchange rates in interstate commerce. There was no national money standard.

During the Revolutionary War the Continental Congress issued paper currency called "continentals" to pay for the war effort. Continentals were accepted at face value to start with, but since the fledgling country needed more and more money to fight the war, the Continental Congress ultimately ended up printing 241 million continentals with only the smallest fraction of backing in gold or silver. The result was exactly the same as it always is: massive inflation. In the 15-year stretch between 1775 and 1790 the continental plummeted from full face value to only 1 percent of face value.

There was an urgent need for the devaluing of the currency to be addressed, a job that fell to the new federal government. The problem was that the United States government was broke. Immediately after the war, the rise of the federal government was met with a great deal of scrutiny. It represented a possible move to consolidate the kind of political power that the colonists had just sacrificed their loved ones to bring to an end. But since the new nation was growing rapidly, some form of national standard was inevitable.

The first attempt at setting up a central bank came with the formation of the privately held Bank of North America. Robert Morris, the Superintendent of Finance of the fledgling federal government, obtained large quantities of gold and silver through loans from France and the Netherlands. Morris then issued new paper currency backed by this supply of precious metal and opened the Bank of North America in January of 1782.

It's important to note here that a loan at the national level is not only a monetary obligation; it requires that certain loyalties be extended toward the lender. The actions of the bank were immediately met with howls of protest over alarming foreign influence, unfair competition against the less corrupt state banks, and the creation of "fictitious credit." The charges of creating "fictitious credit" were true, of course, since that is the actual purpose behind fractional reserve banking. The charter of the Bank of North America was repealed by the Pennsylvania Legislature in 1785.

The next attempt at setting up a central bank came with the formation of another privately held company. The First Bank of the United States was granted a 20-year charter by the federal government in 1791 at the behest of 36-year old Alexander Hamilton, the first secretary of the treasury. Hamilton believed that the establishment of a central bank was necessary as a means to stabilize the new country's currency and improve its credit standing to conduct international commerce.

The bank was to be capitalized with a ten-million dollar investment. Eight million was to be raised by private investors, and two million was to be provided by the United States government, to secure its share of the bank. But since the federal government didn't have two million dollars to invest, the bank did what central banks always do. It created paper money out of nothing and loaned it to the United States government so it could purchase its share.

The founding bankers insisted upon the participation of the government so the bank would appear to be an agency of government. However, many people became frightened as the bank's directors demonstrated the power they possessed to influence the affairs of both government and the economy.

The Congress decided not to renew the bank's charter in 1811. Naturally, chaos ensued, leading to the War of 1812. History tells us that the primary cause of the war was the impressment of American sailors by the British Navy on the high seas. Taking sailors from foreign ships had been instituted as a quick way to increase the number of experienced sailors in the British Navy to help fight their kingdom's war against Napoleon. But the British had already rescinded this policy before the war began. It is much more likely that the colonial Grey Men intended to profit from the conflict and use it to realign the new nation against the "threat from outside our borders" while they quietly restructured the domestic financial system.

> *"If the American people ever allow private banks to control the issuance of their currency, first by inflation and then by deflation, the banks and corporations that will grow up around them will deprive the people of all their property until their children will wake up homeless on the continent their fathers conquered."*
>
> —Thomas Jefferson

The pattern begins to emerge. The forces in play leading up to the Revolutionary War were financial in nature. The forces behind the War of 1812 were also financial in nature. A careful study of the conditions preceding each and every war fought by the United States shows that the underlying discontent leading to war was financial in nature.

The next attempt at setting up a central bank came with the formation of the Second Bank of the United States in 1817. This privately held bank was created with another 20-year charter. After the closing of the First Bank of the United States, the individual state banks had printed too much currency to pay down the expenses of the War of 1812. The resulting inflation had reduced the credit and borrowing status of the United States to its lowest level since the Revolutionary War.

The Second Bank functioned as a clearinghouse; it held large quantities of other banks' notes in reserve. This made it powerful. If you were a small local bank and the central bank held more of your bank notes than the amount of the fractional reserve your bank had on hand, the central bank could put your local bank out of business just by showing up and demanding to exchange those notes for gold.

In theory, the purpose of this vast power was to regulate the lesser banks in the system and prevent them from over-extending themselves. History tells us that wherever vast power is concentrated, you'll also find vast corruption to keep it company. In reality this power was used to coerce those banks to abide by the will of the central bankers in anything they were asked to do. Who would they loan to? Who would they not loan to? Which advance insider information on new lucrative business opportunities would they give to what politician? The central bankers could use this power to enrich themselves while remaining insulated from any newsworthy event they had a hand in creating from behind the scenes.

In 1832, President Andrew Jackson prevented an attempt to renew the charter of the Second Bank of the United States, so the bank went out of business in 1836. President Jackson was the only president under whose administration the national debt was paid in full. At a meeting of the Philadelphia Committee of Citizens in February of 1834 he said, "Gentlemen, I have had men watching you for a long time and I am convinced that you have used the funds of the bank to speculate in the breadstuffs of the country. When you won, you divided the profits amongst you, and when you lost, you charged it to the bank. You tell me that if I take the deposits from the bank and annul its charter I shall ruin ten thousand families. That may be true, gentlemen, but that is your sin! Should I let you go on you will ruin 50 thousand families, and that would be my sin! You are a den of vipers and thieves. I have determined to rout you out, and by the Eternal God, I will rout you out!"

By 1836 the pattern behind the Long Con was apparent in the affairs of the United States. Establish a privately held central bank. Identify the bank as being a part of the federal government. Create fractional reserve money out of nothing and lend it out, especially to the government. Buy favors from government officials to promote business interests. Through inflation, drain away the wealth of all who use the money. When the currency is no longer viable, close the bank and start a new one, or create a war to restart the process. But still, owing to the independent spirit of the American people, the Grey Men were having trouble achieving a solid foothold in the new republic.

The closure of the Second Bank of the United States led to popular distrust of central banks and gave birth to what is known as the "free banking era." Between 1836 and 1863 no central bank was created and banking was left to the states to regulate. Every Tom, Dick, and Harry who could scrape together the $10,000 dollars needed to capitalize a bank could enter the game.

The European branch of the Grey Men saw this trend as a bunch of uneducated people of low birth being allowed to enter an arena they had no right to play in. But the United States was founded on the principle of freedom, and the Grey Men would have to watch the debacle play itself out before they could take control of the American financial system. The states were unable or unwilling to maintain a high standard of scrutiny when chartering a bank and so the average lifespan of banks during this period hovered at around five years. No national currency existed and any bank could issue bank notes of dubious quality. Bribes and back-room dealing were rampant. Lax lending policies, and weak paper currencies that weren't backed by silver or gold, led to an increase in bank failures during this period, which was also known as the "Wildcat Banking Era."

During the Civil War, between 1863 and 1865, a series of bank acts were passed establishing the National Bank System of federally chartered banks. The main goal of this series of acts was to create a single national currency and eradicate the problem of notes from multiple banks circulating at once. A new federally backed currency, known as the "greenback" was issued. A 10 percent tax was levied on all non-federally issued currency, effectively driving it out of the marketplace. The minimum capital required to charter a national bank was now $50,000 dollars, and this requirement removed some of the shakier players from the game.

According to the National Bureau of Economic Research, this period was plagued by a recession every two to three years, and punctuated by the Long Depression of 1873–1879. During the Long Depression, 18,000 businesses went bankrupt, including hundreds of banks, followed by the bankruptcies of ten states.

The National Bank Era culminated in the so-called Panic of 1907. The crisis was triggered by an attempt to corner the copper market in the stock of the United Copper Company. The attempt failed, and the banks who had lent money on the scheme suffered bank runs that quickly spread panic across the nation, as vast numbers of people, fearing a system-wide collapse, hurried to withdraw their deposits. At the time, the United States didn't yet have a central bank that could inject liquidity (by printing money) back into the system.

The crisis was averted by J.P. Morgan who pledged enormous sums of his own capital to stop the rampage. He was successful, but after the panic subsided, those who had the greatest exposure realized that only one man had stood between them and ruin. This realization made a lot of investors very nervous. Senator Nelson W. Aldrich chaired a commission to investigate the crisis and propose a solution. That effort led to the formation of the Federal Reserve System.

There is no better example of the institution of economic slavery than our own Federal Reserve System. In the autumn of 1910, six men attended a highly secret meeting at Jekyll Island in Georgia for the purpose of drafting The Federal Reserve Act, a bill delivering control over the money supply of the United States into the private hands of the Grey Men. The secret meeting was attended by Nelson W. Aldrich, Chairman of the National Monetary Commission and father-in-law to John D. Rockefeller Jr.; Paul M. Warburg, a representative of the Rothschild's and a partner in Kuhn, Loeb and Company; Benjamin Strong, Vice President of J.P. Morgan's Bankers Trust Company; Abraham Piatt Andrew, Assistant Secretary of the U.S. Treasury; Hen-

ry P. Davison, senior partner of the J.P. Morgan Company and Frank
A. Vanderlip, President of the National City Bank of New York. [1]

Owing to the previous misdeeds of the banking industry, the public sentiment at the time was against the formation of a new central bank. The meeting was kept secret because in the words of Frank A. Vanderlip in a February 9, 1933, Saturday Evening Post article, "If it were to be exposed publicly that our particular group had got together and written a banking bill, that bill would have no chance whatever of passage by Congress."[2]

When they presented their scheme to members of Congress, the Jekyll Island attendees were met with howls of protest. Over the next three years they bought newspapers in order to control editorial content and rigged the election of Woodrow Wilson, who was in favor of their scheme to take over the financial system in the United States.

The Grey Men wanted Wilson to win the election of 1912 against the incumbent President William Howard Taft. Wilson's victory was in no way assured. So the Grey Men convinced Teddy Roosevelt to return to politics and throw his hat in the ring as a third party candidate. The Grey Men then contributed large sums to the campaigns of all three candidates with special emphasis on Wilson and Roosevelt. Roosevelt succeeded in splitting the Republican vote and Wilson was elected. Finally the Federal Reserve Act was passed on the evening of December 23, 1913.

These six men represented the houses of the Rockefellers, Morgans, and Rothschilds, who collectively held over one quarter of the entire wealth of people on Planet Earth. They were dissatisfied with the distribution of wealth in the United States and wanted to create a central banking system with themselves in control, allowing them to further consolidate their financial power.

The story of the creation of the Federal Reserve is an interesting one, which illustrates the nefarious motives of the Grey Men. *The Creature from Jekyll Island* by G. Edward Griffin (Amber Media, 1998) provides details on this matter. I haven't analyzed the Federal Reserve Act of 1913, and this book is not intended to be a scholarly treatise on the subject. I trust that this task has been handled articulately by Mr. Griffin and other authors. The truth is that there's a huge difference between what the Grey Men claimed the Federal Reserve would become, and what it has actually become. In the Federal Reserve Act, Congress established two key objectives for the monetary policy of the Federal Reserve:

To maximize employment and to keep prices stable.

This seems like an odd mission statement for the champion of fractional reserve banking to adopt. Economic instability is the main feature of fractional reserve banking. It is unstable by nature. If we wanted economic stability we would have created a monetary system that is entirely backed by precious metal in which fractional reserve lending would be outlawed. The truth is that economic instability opens up vast opportunities to make profit, and that is exactly what the Grey Men want. If you look at the history of the American economy you might conclude that the Federal Reserve System was established to solve the problems created by the Federal Reserve System.

Here is what the Federal Reserve has become. It is a privately held banking cartel that has identified itself as being closely related to the government of the United States. It is not federal, and is not a part of the United States government. In 1913, the Congress of the United States bequeathed to it the sole power to regulate the currency of the United States. However, it was actually established so that the Grey Men could drain away the wealth of the American people, whenever they wanted, through the processes of inflation and deflation.

The Federal Reserve represents the biggest continuing power grab humanity has ever seen or endured. It only took 16 years for the in-flationary/deflationary cycle of the system to reach its first collapse in 1929. The dollar could still be redeemed in gold by the average citizen between 1913 and 1933. The collapse of 1929 was caused by high levels of speculation encouraged by an unstable and highly inflated currency. That unstable currency was created by the Federal Reserve printing more paper dollars than its gold backing could support. Once again, the Grey Men had gone too far.

If that wasn't enough, their next move was to make a play for all the gold. The media of that time, under the ownership of the Grey Men, printed the rationalization that the hard times had caused the "hoard-ing of gold" by the people, stalling economic growth and making the depression worse. Therefore, in 1933 President Franklin D. Roosevelt was forced to issue Executive Order 6012.

"All persons are required to deliver all gold coin, gold bullion, and

*gold certificates now owned by them to a Federal Reserve Bank,
branch, or agency or to any member bank of the Federal Reserve
System on or before May 1, 1933. Failure to do so will be punish-
able by a $10,000-dollar fine and imprisonment or both."*

When brought into an agency of the Federal Reserve the "good
citizen" received $20.67 per ounce of gold paid to them in United
States Federal Reserve Notes that were no longer redeemable in gold.
Fort Knox was built in 1937 to house the growing supply of gold ten-
dered up by the people. By 1949 the supply of gold at the fort peaked
at 701 million ounces. Strangely enough, Executive Order 6012 could
force only citizens of the United States to turn in their gold. International holders of gold backed Federal Reserve Notes were still able to
redeem their notes in gold, which they did en masse.

Who do you suppose possessed the largest stockpiles of gold-
backed Federal Reserve Notes in foreign hands? Why, the top 1 percent of society, which includes the central bankers of the world (otherwise known as the Grey Men), of course! From 1949 to 1971, when
President Nixon finally halted the exchange of paper money for gold,
the amount of gold in Fort Knox declined from 701 million ounces to
an estimated 147 million ounces. Since, in keeping with the practice of
fractional reserve banking, there were many more gold-backed dollars
in circulation than there was gold in the vault, it had been easy to drain
all of the gold away sending it into the vaults of Grey Men located
around the world.

The biggest gold heist in human history had taken place without a peep from the guards. How much gold remains in Fort Knox is a mystery to the general public. It's hard to know exactly who is allowed access to Fort Knox. But since that gold is listed as an asset of the Federal Reserve, one would assume only Federal Reserve staff are allowed in the facility. No one in the public media has been allowed to tour the vaults or report on the status of the gold inventory since 1971. It should be noted that people don't keep big secrets for no reason. They keep big secrets because they have something to hide.

According to the National Bureau of Economic research there have been 47 recessions and at least five depressions in the United States since 1790. In my view all of them have been caused by the practice of fractional reserve banking and the manipulation of markets by those in control of the money supply. It doesn't take a rocket scientist to ask the questions: Who are the people that have been clamoring for the right to set up and run a private central bank? Why are they so interested in being in control of one? Are they on a humanitarian mission to save the population from having to go to war?

That can't be it, as we are always at war.

Do they want to run a private central bank to stabilize the economy so we can avoid the boom and bust of recession?

Clearly that can't be the reason, since there have been 47 recessions since 1790.

Could it be that they want to use the power of the Federal Reserve to prevent depression?

That can't be the reason, since we experienced the worst depression in human history only 16 years after the Fed was established, and we're on the brink of another one at the time of this writing.

Could it be that they want to prevent inflation?

Clearly that can't be the reason since inflation has been a reality, declared or not, since the birth of our nation.

What is the reason they are so interested in establishing private central banks in any country that will let them?

The reason the Grey Men want to set up and run a private central bank that masquerades as a governmental institution is because they can use it to magnify their fortunes by obscene amounts in so many ways that the prospect of it defies the imagination of the common citizen. Does the magnification of their fortunes result from the creation of real wealth? No, it results from the confiscation of real wealth from all those who produce it without regard to race, creed, color, or political persuasion. It doesn't matter whether the country their banks operate in is philosophically communist, socialist, or capitalist. All that matters is that they have a monetary system. Allowing these people to run the economy of our country has been like giving the keys of a giant candy store to a fat little kid who doesn't possess the powers of deduction required to keep from destroying himself. It's like throwing gasoline on a fire to put it out.

Where are we today? We are clearly in Phase 5 of the Long Con. The gold has been stolen. The currency of the United States is not backed by precious metal. The cycle of inflation and deflation has been run too many times. The paper money agreement is about to break.

The currency of the United States is backed only by the full faith and credit of the United States Government. In the final analysis, the full faith and credit of the U.S. Government is represented by the tax revenues it collects from U.S. taxpayers. What do the taxes you pay represent? Your labor. Every dollar the Federal Reserve directs the government to print (or create digitally) that they send out in foreign aid, or monetary assistance to multinational corporations, is an IOU with your name and the names of your children and grandchildren printed on it.

Every fractional reserve dollar printed makes every other fractional reserve dollar in existence worth less. Every fractional reserve dollar printed is a claim upon your future labor. Prices are escalating and the people are growing restless. The king has not yet declared price controls or martial law, but the currency is nearly destroyed. The threat from without was introduced on 9-11 and will come to a head if we go into an armed conflict with Iran or China.

The Federal Reserve candidly admits that its economic policies are immune from interference or direction by the government of the United States. That means we, the people, have no say in the extent to which our future is encumbered. The Grey Men have been totally successful in casting their spell of fear and bondage over the common people of the United States. The Long Con has not only been played out in America, it is the very substance of American history.

But this time it's different. The Long Con has also been played out in the rest of the developed world at the same time. One can only imagine what lies ahead. The spiritual task before us is to make sure the people wake up.

Chapter 23

The Endgame

The Grey Men have already achieved a certain level of control over the lives of common people everywhere. They have accomplished that control by manipulating the paper money of the world from its invention to the present. Anyone who participates in a debt-based monetary system is to one degree or another enslaved by it. Who do you know in the United States today that lives without debt or reliance on social security, which now represents future debt? Those lucky ones are very few in number.

The Grey Men have managed to achieve a high level of control over the affairs of humanity without even declaring themselves as the dictators who pull the strings. History has shown us that dictators are often deposed. If one does not know that one is being oppressed by a group of dictators, much less who they are, who is there to oppose?

The Grey Men have been exerting their influence on international banking standards as they have evolved over the last five centuries. They've been at the control panel, carefully stewarding the evolution of the modern-day banking system to make sure it produces what they want. What they want is more power and more control over a larger share of everything that exists. When do we arrive at the point where enough is finally enough?

I can only suppose that there comes a point beyond which one has all the mansions, jet planes, and yachts one could wish for. After that, what is the measure of success? Is it how many companies one owns? Is it one's position on the list of the wealthiest individuals in the world? Does one go on to own a country or to be the first ones to conquer an entire planet? What is the endgame here?

I submit to you that the most powerful thing you can do is to heal yourself. We covered that technology in Part II. The second most powerful thing you can do is to help heal others. In order to help heal others, it is necessary to understand the conditions under which we are all living. One way to predict what the endgame we are faced with might be is to project current economic trends to their logical conclusion to see where those trends might take us.

Trends in the Money Supply

The precious metal backing of most modern currencies has been removed, making it easier for the Grey Men to create economic instability in any location they wish. The Grey Men have learned that creating inflation in one location and deflation in another causes pressure for resources to move. They have learned how to position themselves to take advantage of that movement. They have used this mechanism to assume control of nearly every legitimate area of commercial endeavor. The trend in management of the world's money supply is toward central control.

Trends in Agriculture

The trend in agriculture is away from the family farm and toward agribusiness on a large scale. Traditionally a family farmer, who also makes the land his home, will husband that land carefully, balancing the input of naturally based nutrients with the extraction of commercial crops in an effort to maintain the long-term viability of the land for the benefit of future generations. On the other hand, agribusiness wants to maximize short-term yields through the application of chemical fertilizers and pesticides, thus polluting the environment and yielding crops that are inferior in nutrients.

Companies built on the same model as Monsanto tout the benefits of the monoculture of vast tracts of land, using genetically modified seed that must be purchased from the patent holder every year. Farmers that use these products are prohibited from harvesting their own seed stocks, placing control in the hands of the Grey Men who decide what the price for being able to continue farming will be. The trend in agriculture is toward central control.

Trends in Food Production

We used to grow our food. Now we manufacture it. The average meal is manufactured somewhere else and travels 1,500 miles to reach our tables. Most fruits and vegetables are produced on factory farms that practice monoculture. Most meats come from highly mechanized feed lot-style operations that are carefully located away from public view.

Processed foods are manufactured in highly automated factories by a handful of companies that now have control over the majority of the food supply. The trend in food production is toward central control.

Trends in Manufacturing

A huge percentage of the world's manufacturing capability has been moved from the developed countries to the underdeveloped countries to take advantage of cheap labor. This is nowhere more apparent than in the United States. It is increasingly more difficult to find any typical household goods made in the U.S.A. We've literally sold our heavy industries and a great deal of our light manufacturing capability to corporations in third world nations.

The price of goods has taken precedence over the quality of goods. So now we are surrounded by a proliferation of cheap goods that we are having more and more difficulty buying, since we have lost those manufacturing jobs that provided the income with which to buy them. In the final analysis, when we combine the cost of lost income from jobs exported over seas with the price of cheap goods manufactured on foreign soil, we will find those cheap goods to be the most expensive poor quality goods in history. The trend in manufacturing is toward central control.

Trends in Retail Sales

Our declining economy has forced us to consider that price is more important than quality. More and more people are being forced to shop at Wal-Mart, Costco, and other "big box" retailers. When you go into a Wal-Mart superstore, your first impression is that they have everything you could ever want. Seeing so many different kinds of goods in one location is truly impressive. However, a careful study of what is available from the mega-source stores shows that there is actually less variety being sold on the open market than used to be available from the mom and pop stores of the 20th century.

The reality is that the super store business model actually restricts the number of choices you have in the marketplace by eliminating its competition. The big box retailer decides what will and won't be placed on the shelves. The trend in retail sales is toward central control.

Trends in Health Care

The health care system in the United States is now run by insurance clerks who decide which treatments the system will or won't pay for. The people who make these decisions do so from a cubicle somewhere else according to a script devised by medical accountants. They have no relationship with the patient's actual needs, and no contact with the patient's actual condition. This myopic view of health care restricts the types of treatment available to the western medical modalities recognized by the insurance companies. Those treatment options just happen to coincide with the business plans of the largest pharmaceutical companies.

You may have noticed that when you go to the doctor you almost always come away with a prescription to buy. The doctor is not telling you how you can take your healing into your own hands. The doctor is telling you that your healing is out of your hands and comes from those in authority who occupy a higher station in life than you do.

You must apply for and be granted access to the drugs that you saw advertised in the magazine you were reading (thereby having become convinced they're what you need). We have come to the point where health care is so expensive, any family member with a serious health issue that is not participating in the system will likely end up filing for bankruptcy. This system of health care central control is so invasive that the main solution being considered is to bring it all under the central control of the government.

We can go on and on, but I think you are beginning to see that the trend in all major areas of commerce is toward more and more central control by the people who control the one system upon which all other systems depend: The monetary system. This is all coming to the forefront on the heels of the dissolution of the Soviet Union where it was proved beyond a shadow of doubt that central control of society doesn't really work! The one big lesson of the Soviet experiment was that they had devised a system that ignored one hugely important premise, which I repeat here:

The personal expression of free will is a basic human instinct that affects the core motivations of all human beings.

It should be clear by now that celebrating each human's ability to live in a natural state of liberty, enjoying the maximum opportunity to express free will, is a universal success strategy. Are we willing to bet that the synergies that flow from removing the spell cast over humanity, and empowering the average world citizen to create more fully, will manifest a better world for all?

Conversely, do we not see that continuing to enslave the people of the world under any system of central control for the short-term profit of a few individuals is a doomsday strategy that is not sustainable? If the current trend toward the centralized top-down management of all humans continues, the logical conclusion will be a one-world government and a one-world currency under the control of the Grey Men. Though they would like you to think otherwise, these are people who are not necessarily more evolved than you are. They are actually standing in the way of evolution.

Like it or not, humanity is slowly starting to realize that we are all one species here doing the same things for the same reasons. We all came from the same source consciousness and we are all participating in the expansion of that consciousness. If we are to design a new reality that makes the problems of the old reality obsolete, a one-world currency could be a quantum leap forward in human evolution. It could make it easier to transact global business. It could contribute to a higher level of economic stability. It might have the potential to ensure that all peoples have an increased opportunity to enjoy life, liberty, and the pursuit of happiness by ending the manipulation and control that is the hallmark of the current situation. However, such an outcome would depend entirely on shifting the paradigm behind the system.

Under the existing paradigm, a one-world currency is fraught with danger. The current trend in banking is to make all currency obsolete in favor of a system of electronic credits. Can you imagine the abuses that could take place under such a system? The first thing that would need to be put in place for such a system to exist is an electronic identity for every man, woman, and child on the planet. Your electronic ID would access a data base containing your complete financial profile to determine whether or not you were eligible to complete the transaction you were contemplating. The system would not only record all your shopping trends, it would also record your physical whereabouts. We need to do everything in our power to make sure this does not come to pass.

If you were say, a political dissident with a philosophy opposed to the current control structure, one key stroke could prevent you from being able to engage in any commerce whatsoever. You wouldn't be able to buy gas or food or clothing.

No one on the planet should have that kind of power over others. On the other hand, a physical one-world currency that was not born of the inflationary influences of fractional reserve banking, and that did not serve as an instrument for the manipulation and enslavement of the people, could be an enormous improvement. A physical currency that would maintain both stability and *the crucial anonymity of its users* could usher in a whole new era of real economic stability and sustainable living systems.

One world government is another story. One of the things that
irritate Montanans is the idea that the city dwellers in Washington
should have the right to create one-size-fits-all legislation that applies
equally to all regions of the country. If you spend any time with rural
people you will find that they share a common aversion for the traffic
congestion, excessive noise, air pollution, overcrowding, and crime of
urban environments. The country folk who enjoy watching the sun
rise and set over the horizon, instead of the neighbor's roof, don't want
to trade the ability to make a sumptuous meal from fresh vegetables
out of the garden for the fact that Pizza Hut is just down the street.

The differences between urban dwellers and rural people offer
even more contrast than do regional differences from north to south
or east to west. There are a lot fewer people in rural settings. Rural
people enjoy a different level of freedom than the average city dweller.
Accordingly, government should be local and regional, reflecting the
desires of the people who live in any particular locale. One size does
not fit all.

The idea preached by the founding fathers of this country was that
the federal government should only have certain powers to protect
the whole and the powers that are not specifically given to the federal
government are reserved for the states.

Wouldn't it be nice if the governments of the world could agree
on a similar paradigm for governing that is really of the people, by the
people, and for the people?

That is what we say we are about in the United States. However, we
have demonstrated to the world at large that we do not really honor

our own philosophy. At the highest level of our society, the movers and shakers in business and politics subscribe to a paradigm of manipulation and enslavement in order to confiscate the wealth of the many for the aggrandizement of the few. It has not been my intent in the writing of Part IV to identify the Grey Men as the enemy. Rather it has been my intent to identify the paradigm that they live under as one that should be replaced as quickly as possible by a new cosmology. That replacement will occur as we each examine what is in our hearts and minds, and make suitable changes based upon universal principles.

How do I reconcile the idea presented in Chapter 10, that we are each responsible for generating our own experience, with the notion that the Grey Men have cast a spell over us all? Simply put, we have allowed a spell to be cast upon us by not waking up and seeing it for what it is. We have accepted what we've been told about life and we have participated in the illusion. If we fail to act, someone out there is willing to act on our behalf without declaring that they are doing so.

The quickening is going to manifest the logical extension of everyone's individual belief system. The Global Financial Crisis is evidence that this is happening. If your religious or economic belief system is fear based, it will not be sustainable in the higher ages that are coming. If it is not sustainable, it will eventually burn itself out. I wrote about the Grey Men to show you how we've accepted the role of slavery we've been playing.

Because of the quickening, we have an enormous opportunity to bring on the higher ages before they are scheduled to appear. If you understand the implications of the Yugas, you will see that our new cosmology for the 21st century has indeed arrived ahead of schedule. Judging by what is still going on around the world it doesn't appear that we should have been able to wake up quite this soon. There will still be an enormous number of people who like the existing system the way it is. They don't want to change themselves and they don't want you to change either.

We need to make a united and non-violent stand against the forces of central control by initiating success stories that demonstrate the strength behind decentralization. Big government is not the answer. If we can bring ourselves to believe that there is more potential available from investing in the creative abilities of all people than there is in controlling and enslaving them for the benefit of a few, then we truly stand on the threshold of a quantum leap in the development of human consciousness.

Chapter 24

Decentralization

The real motivation behind the idea of central control is the paradigm of exclusivity. Those that wish to establish such a system will tell you that the ideal they are pursuing is equality for all. They'll say that the common people are not sufficiently educated to understand how to make the intelligent decisions necessary to fulfill their own desire to enjoy a healthy and happy life. Therefore, they'll say, those people must be supervised by a benevolent class of more intelligent people who know better than they do what is good for them.

The real motivation is equality for the lower tiers of society while the privileged overseers enjoy a more affluent lifestyle suitable to their elevated station in the culture. In other words, they want you to pay them to decide your future for you. Certainly in this country there is a demographic group of people who are less capable of running their lives than the average person. They may lack education and live under the psychological and physical strains of extreme generational poverty. They may lack access to resources, and be stuck struggling in survival mode. Our motivation should be to help educate those friends and neighbors of ours to learn to run their own lives well outside of government, rather than to create dependence on a system of central control that reduces everyone to the lowest common denominator. The belief system of exclusivity does not support each person's right to express his or her free will to create.

What really drives this movement is the desire by certain people to be admitted to the exclusive class of overseers and to enjoy its benefits. Remember the quote by Baron Amschel Mayer Rothschild: "The few who can understand this system will either be so interested in it profits or so dependent on its favors that there will be no opposition from that class, while on the other hand, that great body of people mentally incapable of comprehending the tremendous advantage that capital derives from the system, will bear its burden without complaint and perhaps without even suspecting that the system is inimical to their interests."

The real meaning of this passage is that once the system of central monetary control is established in a particular country, there will be a group who are on the inside and whom will enjoy its profits. They will manage the other group, on the outside, which will be enslaved to one degree or another by the system itself, perhaps without even being aware of it.

In 1805 it may have been true that the great body of people could not comprehend this system. But if the real story is exposed today, the great majority of people in the United States and Europe are quite capable of understanding what has been done and why they are experiencing the debt related hardships that have become a normal part of everyday life. In the developing nations of Africa, Asia, and South America, where education systems have lagged behind the standards of industrial nations, and where the culture of debt has not yet taken hold, comprehension may not be so immediate.

In the United States, government that represents the will of the people has been purchased by big business. Do you still have personal contact with your representatives in a meaningful way? Probably not. Big business has stepped between you and government and has corrupted the process and its participants to its own interests.

The message here is that central control and big government are bad for you. Our founding fathers were well aware of all these issues. That is why this country was created. This country was founded to celebrate the inalienable right to life, liberty, and the pursuit of happiness provided by the Creative Principle, not by a group who wants to charge you a fee for it.

Both of our main political parties have demonstrated their total commitment to remaining in the membership of the ruling class. They cater to the whims of big business whether or not those whims benefit the people. They've exempted themselves from the one-size-fits-all health care program they want to impose upon you. They receive lifelong pensions for merely being elected. They are allowed to spend money they don't have. They have printed vast sums of fractional reserve money, for which you will eventually pay the bill, and they've spent that money around the world without your knowledge or consent. When they leave Congress, they accept lucrative positions lobbying for the big corporations they have served for so many years.

Phase 3 of the Long Con, in which the government is bought by the bankers, was accomplished in the United States a long time ago. You may be hip deep in the national debate as to whether it should be the Republicans or the Democrats who rule next, but make no mistake, whether they understand it or not, they all work for the same group and that group is not we the people.

Government of the people, by the people, and for the people is extinct in the United States and has been for some time. The national political debate rages on and yet nothing of substance really changes. Our currency and our culture are still being systematically destroyed by the operators of the secret monetary system. What is needed within the political process is a charismatic leader and a group of committed statesmen, who understand the real truth behind what has been created and those who have the courage and popular support to stand against the secret system and effect real change. I believe the popular support would be there to greet such people, but as of this writing no such group has stepped forward to be recognized by the populace.

What is there to do? First and foremost we should halt the trend toward central control. In order to return government to the purpose of being of the people, by the people, and for the people, it will be necessary to return the main functions of government back to local control. Nature shows us that there is strength in diversity. We are just beginning to enter an era when people will acknowledge that we are all here doing the same thing for the same reasons. We are all engaged in the process of expanding our consciousness in one degree or another. Having the realization that we are all connected doesn't mean we must all become alike. We can respect and even celebrate our differences if we can all agree that the best way to create a successful world is:

As much as possible, do no harm to others.

Rather than concentrating on centralizing control we should be concentrating on decentralizing control. We should recognize that the intent of the founding fathers was for the federal government to provide a protective shell for the whole and no more. To centralize control is to create massive opportunities for corruption, which is exactly what surrounds us. Central control does not work!

We need to re-establish our culture from the bottom up. We need to bring our own survival capabilities home. We should grow our own food locally. We should make our own goods locally. We need to re-establish the butcher, the baker, and the candlestick maker as necessary components of a solid local culture. Our children should be able to go down the street and see how everything we use in daily life is made within the local community. We are one of the most technically advanced countries in the world. There is no reason we can't enjoy a proliferation of very high-tech small businesses and cottage industries.

Some people will say, "We can't afford to buy American goods." If you look at what the trend toward centralization has done to our country you could easily conclude, we can't afford not to buy American. We need to make the goods we consume on a daily basis and adjust to the economy required to do so. We need to bring our jobs back home.

We know there are great changes coming in our monetary system. We should put a stop to fractional reserve lending and demand that the federal government live within its means. We should demand that whatever new monetary system is put in place that one crucial feature is maintained—everyone must be able to maintain complete anonymity when buying what they need, if they so choose.

Those who love the existing system will say that an anonymous currency creates more opportunities for people to cheat on their taxes. Then so be it. Would we be better off living with a small percentage of people who cheat on their taxes or with a system of electronic credits that allows the Grey Men to monitor every transaction we make? That would truly be giving the fox the keys to the hen house.

One world government is a bad idea. Even national government with sweeping powers to manage the everyday affairs of all the people is a bad idea. If it has proved nothing else, our government has shown us all, without a doubt, that its members don't have the backbone to resist the temptation to raid the cookie jar and overspend the piggy-bank, creating cushy lives for themselves at our expense. They not only want this paradigm to continue, they want it to increase.

The system of the Grey Men is already the most invasive thing on the planet. They control the money supply of the developed world. They own the Wall Street system that is used to re-distribute wealth. They own the military/industrial complex that is used to destroy infrastructure. They own the multinational construction companies that are used to rebuild that infrastructure. They own the food production and distribution system. They own the health care companies and insurance companies. They own energy companies in all their various forms. They have also mired us in mountains of debt we can never hope to repay. The people are growing restless. They are becoming tired of working to pay these bills. Will we allow the Grey Men to take an even higher level of control over us?

There are two things that stand between us and total subjugation. The first thing is that ten generations of Americans have been raised with a higher level of freedom and independence than any culture on Earth. That sense of liberty will not be extinguished easily. The second thing is the fact that the American people are educated. We are quite capable of understanding the spell that has been cast upon us. We are quite capable of designing and building a future that makes the problems of the past obsolete.

Sometimes the price of liberty is high. It is not yet time to give up your right to defend yourself. You may still be called upon to defend your liberty. How does that live in the same philosophy with "do no harm to others," you ask? We are only 300 years into the Dwapara Yuga. If you have read this far you know that there are still very powerful people on the planet that want to increase the level to which you are managed and enslaved. Many of them live within our own country.

The fact that the Grey Men have confiscated 80 percent of all the wealth on the planet should be enough evidence to convince you that they do not recognize your right to self-determination. There are a lot of people who will vigorously oppose the decentralization of government. They like it the way it is and will not want it to be rearranged. There may come a point in history where it will be necessary to oppose tyranny once again. In the meantime, we can still live a balanced life by pledging internally to do no harm to others. We can still live a balanced life by being the peaceful warriors who are going about the business of creating a new reality that makes the old one obsolete.

Part V

The Planetary Free Will Experiment

Chapter 25

Extinction through Overspecialization?

At an annual meeting of the American Association for the Advancement of Science in the early 1950s, two complementary research papers were presented. Unbeknownst to each other, two scholars attending the conference presented separate papers, one in biology and one in anthropology. The biological paper investigated the histories of biological species that had become extinct. The anthropological paper reviewed the histories of human tribal cultures that had become extinct.

Each scholar, unaware of the efforts of the other, had come to the same conclusion. They each concluded that the cause of extinction for both biological species and human tribal cultures was a phenomenon called overspecialization. The tribes and species in question had become so specialized in doing things only one way that they lost the ability to adapt to changing circumstances. When unforeseen changes occurred, they were unable to react, and thus they had perished.

In his groundbreaking book, *Operating Manual for Spaceship Earth* (1978 E.P. Dutton), R. Buckminster Fuller gives the following example of the process of extinction through overspecialization: "There once was a type of bird that lived on a special variety of micro-marine life. Flying around, these birds gradually discovered that there were certain places in which that particular marine life tended to pocket—in the marshes along certain ocean shores of certain lands. So, instead of flying aimlessly for chance of finding that marine life, they went to where it was concentrated in bayside marshes. After a while, the water began to recede in the marshes, because the Earth's polar ice cap was beginning to increase.

"Only the birds with very long beaks could reach deeply enough in the marsh holes to get at the marine life. The unfed short-billed birds died off. This left only the long-beakers. When the bird's inborn drive to reproduce occurred there were only other long-beakers surviving with whom to breed. This concentrated their long-beak genes. So, with continually receding waters and generation to generation inbreeding, longer- and longer-beaked birds were produced. The waters kept receding, and the beaks of successive generations of the birds grew bigger and bigger. The long-beakers seemed to be prospering when all at once there was a great fire in the marshes. It was discovered that because their beaks had become so heavy these birds could no longer fly. They could not escape the flames by flying out of the marsh. Waddling on their legs they were too slow to escape, and so they perished. This is typical of the way in which extinction occurs—through overspecialization."

Have we generated a modern society in which we have become over-specialized? The answer to that question is absolutely. Yes. Our monetary system itself is a form of overspecialization. We have set up our society so that we must use money to buy nearly everything we own or consume. Other mediums of exchange, such as barter, are an insignificant percentage of total commerce. When currencies are destroyed, as they frequently are under a fractional reserve banking system, people are threatened. In the same way that the changing circumstance of receding waters made it difficult for the long beakers to obtain sustenance, the difficulty of gathering money during a recession or depression makes it harder to acquire the necessities of life. However, while our monetary system is at the heart of our troubles, there are other ways in which we have become over-specialized.

When I was a boy in the early 1950s our family only had one car and dad drove it to work, so wherever the rest of us had to go we went on foot. When it was time to buy groceries I would accompany my mother to the local grocery store. My job was to pull the little wire basket on wheels that we used to transport our groceries back to the house. This wasn't really an inconvenience since there was a little mom and pop grocery store around the corner from our house. The post-World War Two economic boom was just getting started and the two-car family wasn't yet common.

The design of our neighborhood provided for us to do what we needed to do without getting into an automobile. We would pass other mothers coming home from shopping on foot and would stop to say hello and catch up on the news. Sometimes we would get together with other mothers and their kids to go shopping on foot. We had plenty of time. There was no sense of being deprived of the convenience of driving everywhere since we had never known that convenience. My mom didn't learn how to drive until my 16th birthday when we both acquired our first driver's licenses in the same year. As the post-war boom began to take hold, we all fell in love with the automobile. Everybody looked forward to the yearly introduction of the new models, each one more jet-like than the last. The people of the United States had saved the world from the tyranny of the Nazis and Imperial Japan. The right to own a car was seen as a reward for being part of the best country in the world.

As the post-war economic boom accelerated, the development of real estate also accelerated. Up to that point housing was mostly constructed, one home at a time, by small local contractors. Then the Grey Men opened up the cheap credit window to finance the American Dream and real estate development took off. The more we master-planned our communities, the more we designed them around the automobile. We loved our cars. We loved being able to go anywhere we wanted, whenever we wanted. Gasoline was 25 cents per gallon, insurance was cheap. My first car had a six-gallon gas tank and I could drive for a week for a dollar and 50 cents.

The more real estate we developed, the more we had to rely on our cars. The little mom and pop stores that lined Main Street gave way to centrally located shopping centers. As our main streets and the buildings that lined them aged, we moved farther away from town centers into the glorious new world of the suburbs. The automobile gave us the freedom we all craved. The streets weren't crowded, traffic jams were rare. Gradually three things happened that we didn't really notice.

In the distant past, local families had economic relationships with other local families. Families would get together at harvest time and help each other bring in the crops. They would trade milk and meat and eggs for handcrafted goods and services. They made their own clothing and furniture. They built each other's homes. If they couldn't afford it, they didn't buy it. There was little outstanding debt. They were related to each other by the fruits of their labor. Society had a stronger fabric of relationship under it. As the miracle of cheap credit took hold, they were able to buy more and more goods. As individual debt mounted, so a sense of needing to run a little faster to keep up took hold. Life was good though, for the most part we were able to pay the bill.

As time went on, the first thing we didn't notice (or didn't think was a problem yet) was that we were becoming less related to our neighbors and more related to the financial system of the Grey Men. There was no longer a need to go through the hassle of trading with your neighbors when you could go to the supermarket and purchase a wonderful variety of things never imagined before with your cheap money.

Being modern was the driving force behind our society. If you look at the advertisements of the day you see the Betty Crocker moms in their high heels and summer dresses gaily executing their wifely duties with their new space-age vacuum cleaners and washing machines. Slowly the character of our relationships with our neighbors changed. We no longer gathered together because we had been related family to family for generations. We sorted ourselves into the landscape based upon what we could afford.

The more we borrowed to finance the American Dream, the less we were related to each other and the more we became related to the financial system. The more we borrowed to finance the American Dream, the more our incomes were seen as the growing ability to service greater and greater amounts of debt. In retrospect, the more we borrowed, the weaker the underlying social fabric of our culture became. This entire phenomenon was based upon the optimism of being the greatest country in the world. We were growing, nothing could stop us, and it was going to go on forever until there was a helicopter in every garage.

As our general self-reliance declined, our reliance on the dollar increased. This was considered normal and highly desirable. With the dollar, you could get anything you wanted from someone who specialized in making just that one thing. As kids we all heard the stories our parents told about the Great Depression of the 1930s, how they had to "do without." Living through that period made such an impression on our parents they still practiced some of the frugalities they learned as children.

I remember very clearly the day when the father of one of my friends was driving us to school in his standard shift car. He declared that the object of good driving was to get where you wanted to go without having to touch the brakes since brake linings were expensive. We made fun of him and told him he should get with the program. This was the '60s and we were on our way to the moon! We wanted to peel rubber with no concern over the cost of tires—and we did!

Without knowing or understanding what we were doing, we all became more and more enslaved to the culture of debt. Being a specialist was looked upon as a highly enviable position to be in. The technology of the '60s was so complex one could never have enough time to specialize in more than one field. Without knowing or understanding what we were doing, we were well on our way to becoming over-specialized in every area of modern culture. It was all predicated on the concept that growth was going to continue and circumstances were not going to change.

Then the second thing that we didn't notice began to happen. As the real estate boom took hold, everything we built was designed for the automobile. We created massive residential areas that were so huge one had to drive a car just to escape the suburbs to go do any kind of business. We separated the functions of society by placing commercial activity away from residential areas. We centralized our shopping and business functions. We created more and more roads and superhighways and parking lots. Then, we sorted ourselves into the landscape based upon how far we could conveniently travel in our cars. It became common to live on one side of town and work on the other. No one questioned why we were all crossing town every day in the first place. In the beginning it was considered to be a pleasure. We gave birth to the commute, the traffic jam, and massive air pollution. We designed more and more luxurious cars to make the time we spent trapped in our cars more enjoyable. We invented automatic transmissions, air conditioning, satellite radio, and heated seats to increase our enjoyment.

As these trends accelerated we slowly began to see we'd become enslaved by them. Driving to work was no longer a privilege to be enjoyed, but a hassle to be endured. We watched in horror as the price of gasoline ate into our ability to pay for the other goods we wanted. Because of the way we designed the built environment, we had to drive to fulfill the daily functions of living. God help us if it ever became so expensive to drive that we would no longer be able to. The oil embargo of 1973 gave Americans a huge scare. We all sat in lines at the gas pump wondering what we would do if we couldn't get gasoline. How would we get to work, how would we go shopping? We had unknowingly become over-specialized in yet another area of our lives.

Then the third thing began to happen. As fractional reserve money poured into the system, the dollar inflated and prices started to escalate as they always do. The same home we bought in the early '60s for $34,000 dollars cost nearly a half million dollars in 2008. During the 1990s the construction industry went into hyper drive and started to account for a larger and larger share of the gross national product (GNP).

Americans were too sophisticated and too successful to continue to do the menial jobs. We brought in third world labor to dig our ditches and install our roofs. Our labor unions pushed the price of semi-skilled labor out of balance for the work that was being done. In an effort to escape the limitations of organized labor and produce cheaper goods with lower prices, our manufacturing companies began to move their operations overseas to take advantage of cheap labor.

At home we had to pay $30 dollars an hour for someone to push a button. Over there we had to pay $30 dollars a month. If we could have made more money exporting the process of building homes we would have done it. The construction boom became the engine of economic growth. General building contractors were popping up everywhere. We didn't mind if that growth was no longer centered in manufacturing. This was the age of information. Since we were so technologically advanced, we would no longer need to actually make anything on our own soil. Those of us who were not in the housing industry would have white collar jobs and obtain whatever we needed with the power of the dollar.

Gradually the profit motive drained away our manufacturing capability. You could no longer go down the street in America and find a place where something was being made that wasn't for the housing industry. People that once worked in factories crossed over into both residential and commercial construction. As our manufacturing force dwindled and our construction labor force increased, once again we had put all our eggs in one basket. Once again we had over-specialized. Then, in the late 1980s, the housing market started to slow down. The politicians were clamoring for the right of all people to own a home. The American Dream should be available to everybody.

The Grey Men realized there was at least one more expansion of the inflation bubble available before the currency collapsed, but they would have to be clever about it. They opened up the credit window with highly innovative subprime financing and we were off to the races one more time. The '90s produced staggering growth and staggering price increases.

The business of buying a home in a good area, making minor improvements, and flipping it right back on the market to take advantage of escalating prices took off. The Grey Men decided to add one more layer to the intrigue and began to bundle subprime mortgages with more solid conventional mortgages in Triple A-rated securities. Those securities were traded around the world. Then circumstances changed as they always do, and the marshes began to dry up.

The dollar had been inflated past its ability to sustain any more growth and it stalled in mid air. The same thing was true of American real estate. Even though we had witnessed a similar phenomenon in Japan during the '90s, no one believed it could ever happen here. Most felt there was a lot more room in the current worldwide bubble for currencies to inflate.

As real estate prices started to sag, the collateral underlying the massive mortgage backed securities market went into retrograde and began to pull down the financial system with it. But having established the fractional reserve banking system all over the world, the Grey Men had indulged in overspecialization, too. As a result, the financial epidemic spread like wildfire around the globe.

More and more people are experiencing firsthand knowledge of the crisis. As real estate prices plummet, their retirement accounts are dwindling, the equity in their homes is disappearing. Many have lost their jobs and cannot pay their bills. Money is harder to come by. In the past the Grey Men could work the Long Con in countries A, B, and C and move to countries D, E, and F to start the process again. This time it is different. The Grey Men have been so successful at spreading their spell around the world that there is no place left to go where they don't meet themselves coming the other way. The "long-beakers" have become "too heavy to fly" and the "marsh" is on fire.

The quickening in the power of manifestation as our solar system moves back toward galactic center is bringing the paradigm of lack and fear to its logical conclusion. The natural extension of the paradigm of lack and fear is self-destruction through the inability to adjust to changing circumstance, otherwise known as extinction through overspecialization. Are we talking about the biological extinction of the Grey Men or the common people? Or are we talking about the extinction of belief systems, financial systems, and governmental systems? The answers to those questions will depend on how we react to what is happening.

Our cosmology for the 21st century tells us that whatever we experience collectively, the feelings we are having about what is happening are completely personal. Harboring fear or blame in our hearts is equivalent to asking the Creative Principle to provide more of the same in our future experience.

The common forms of everyday life are going through a transition. Many of the cultural forms we are so used to are beginning to dissolve around us. Those of us that have come to rely on traditional cultural forms for our sense of stability and well-being may be swept away emotionally by the quickening. In order to survive and prosper, we will need to manifest an inner sense of well-being despite what is happening around us. Remember, we won't necessarily get what we want; we will get "what" we are (meaning, how we choose to think and feel will be returned back to us by the Field).

Unlike the long-beaked birds, we are not stuck with one-speed evolution. We have something else we can use to manifest the future we want. We are creators-in-training and we can create our way into a better life. The first step is to fill our awareness with the understanding that we aren't victims. The world we've manifested as a collective is the result of our own individual use of the creative process.

The universe is still in balance. As we accelerate into Dwapara

Yuga, we are surrounded with opportunities to create a new reality that makes the problems of our old reality obsolete. First we have to wake up and remember that we are the ones who are capable of transformation. We have a duty to work with our children and grandchildren to take our place in the annals of history as the three most creative generations that have ever lived.

Chapter 26

The Paradigm Shift

The world is still a dangerous place. There are legions of people who, when presented with an opportunity to enslave others in order to improve their own economic standing, will gleefully do so while celebrating their good fortune. Unfortunately these people tend to congregate in positions of authority for the specific purpose of acquiring power over other people. They tend to pursue great wealth at the expense of others, giving little thought to what the unintended consequences of their actions might be. Tending to be materialistic in their orientation to life, they measure success in the pursuit of happiness by how much pleasure they feel.

Pleasure is born of the senses. If you have mistaken the pleasure of the senses for true happiness and well-being, then you are likely to feel compelled to acquire objects of greater and greater sense pleasure. The gilded trappings of the world may provide a temporary sensation of pleasure. However, like drug addiction, more and more of the objects that have been identified as creating that pleasure are required to achieve the same result. You will need a new home, a new wife, a new car or a new financial conquest if you want to continue to feel the same level of pleasure, the same buzz of adrenaline and brain chemicals, as you did when you made your first acquisitions.

This mistaken view on the nature of real well being is what drives some people to confiscate all the wealth they can get their hands on.

No matter how much they acquire they're still nagged by the fear that they won't be able to continue satisfying their ever-growing need to experience sense pleasure. This error is also what makes them think they are successful, and that they are doing the right and proper thing. After all, if everyone you meet is envious of your position in life, you must be on the right track.

Joy is born of consciousness which is reflected in the heart. Joy is born through understanding the intent of creation, thereby experiencing the process of expanding one's consciousness. Once carefully attained, the joy of the heart slowly becomes a permanent state of bliss. Bliss is a state of consciousness that requires no outside objects of stimulation to exist. Bliss is a permanent way of being that survives even the body itself and carries the consciousness toward a greater and greater communion with the source energy of the Creative Principle. This truth has been demonstrated by every spiritual master down through the ages, and it is a good thing to hear about. It is an even better thing to discover by the firsthand practice of the ideals that lead to higher consciousness.

I have consumed quite a few pages telling you about the spell that has been cast and about the mindset of those who have done the casting. It is not my purpose to write about the Grey Men to help you identify who the "next enemy" is. Remember, if we express the emotion of being victimized, we are simply directing the Creative Principle to produce more of that in our lives. Nobody that is on a quest for truth and meaning wants to sign up for that. My purpose in describing the way our financial system is structured has been to give you a sense of why things have turned out the way they have, and to suggest ways to respond.

In the current political climate many people are lining up for the privilege of identifying and prosecuting the bad guys. Are we going to have a class war? Are we going to war against a certain kind of person or against a philosophy? There is a tendency to blame the wealthy for the global financial crisis. That blame is not misplaced, but what are you creating if you allow yourself to indulge in it?

It is true that greed has been one factor in creating this situation. It is also true that greed results from a lack of awareness of the real intent behind creation, which is to expand consciousness. Before we roll out the guillotine, shouldn't we take intent into account? Certainly there are those who have practiced the convoluted art of confiscating the real wealth of the common people through financial trickery. Such people are not worthy of the admiration we tend to pile upon them. They may live lives of continuous sense pleasure, but many have shut themselves off from the experience of true joy. You cannot buy true joy upon the suffering of other people. In my view this kind of life is the pitiable result of ignorance of universal principles.

Then there are those who have achieved massive prosperity by creating real value in the lives of millions of people. These true innovators are to be admired. They help to push the evolution of humanity forward by demonstrating the power of the creative process. The computer, the cell phone, the iPod, and the Internet have all made huge contributions to the expansion of human consciousness.

Shouldn't we celebrate prosperity acquired through real creativity?

Then there are those people who have both created real value in the lives of others and have enslaved people at the same time. How do we evaluate them? Do we stand against a class of people or against a belief system? Intent matters. We are all engaged in experiencing the unintended consequences of our actions. If we stand against a class of people, doesn't that create an equal and opposite reaction in the Field that we might not want to sign up for?

Perhaps we should spend the least amount of our time standing against anything and dedicate the largest amount of our time standing for things that make our problems atrophy and fall away. You cannot manifest the world you want by making war on the world you don't want.

Certainly, it's useful to know what has happened and why it has happened. There is nothing wrong with calling a spade a spade. But once we know and understand why things are the way they are, shouldn't we use that information to create a new way of being that makes the old paradigms obsolete? There is indeed a time to stand in opposition to the willful infringement of liberty.

We can do that at the ballot box to a small degree. But we need to do more than that. We need to discuss our understanding with our neighbors until it becomes commonplace. We need to bring about a paradigm shift. In order to achieve that, I believe we need to be more focused on creating the future than we are in preventing the abuses of the past. If we dedicate ourselves to creating paradigm shift in others, we'd better make sure our own houses are in order. Start by examining your own heart. If you find a need there to manipulate others for your own gain, you know what you need to do. Shift your thoughts and emotions.

Who among us is qualified to evaluate the true intent of another person? If we decide to elevate ourselves to the level of judge and jury don't we expose ourselves to the dangers of hypocrisy? Don't we become what we want to have disappear? Perhaps we should think of the Grey Men more as a belief system than a class of people.

We need to keep investigating what has gone on, and what is still going on, that we do not want. We need to keep shining the light of truth on the practice of enslaving others no matter who is doing it. We may actually be surprised how quickly the old paradigm will atrophy and fall away. Acceleration is a difficult thing to plot on a chart. We know the affairs of humanity are accelerating toward a logical conclusion. If we are diligent, the paradigm shift that needs to happen could occur in a single generation.

Chapter 27

The Intent of Creation

The spiral nebula on the cover of this book is not our Milky Way Galaxy. It is a galaxy that is very similar to ours. We have not yet been able to place a spacecraft far enough outside of our own galaxy to look back and photograph it. In fact, after 35 years of traveling through space, the Voyager 1 Spacecraft is only now crossing the outer boundary of our solar system. Voyager 1 is estimated to be 11 billion miles from the Sun, and has the distinction of being the manmade object that's farthest away from the Earth.

The Kepler Space Telescope was launched in 2009, with the mission to discover planets, like ours, that may be capable of supporting biological life. The latest information coming in from the Kepler is there may be as many as 100 billion exoplanets, orbiting around their respective stars, within the boundaries of the Milky Way Galaxy. An exoplanet is defined as a planet outside the bounds of our own solar system. The Kepler Space Telescope watches a patch of sky containing over 100,000 stars at a time. It waits for slight dips in the brightness of those stars. The dips occur when an exoplanet passes in front of its parent star, blocking a small fraction of starlight from our view.

Our astronomers are able to analyze that data and tell the mass of the star, the mass of the planet orbiting that star and how far they are from the Earth. The data coming in suggests that almost all of those star systems have at least one exoplanet and most have multiple exoplanets orbiting the parent star. It is now estimated that two thirds of the star systems in the Milky Way Galaxy include planets similar in size and mass to our Earth. Of that number, we don't yet know how many Earth-sized planets occupy orbits that are the right distance from their central suns to allow the formation of biology. The number could be as high as 500 million. The odds that life on Earth is alone in the galaxy are diminishing day by day.

In Chapter 7, we discussed how the Earth is home to a hierarchy of individual units of self-aware consciousness ranging from the smallest microbes, to human beings, and even the planet itself. Each one of those organisms is busy collecting experience, so we can say that each unit of self-aware consciousness is expanding its awareness day by day. In fact, it is difficult, if not impossible, for consciousness to inhabit biology without expanding its body of knowledge.

If we pull back far enough to see planet Earth floating in space we can describe our planet as being a container dedicated to the expansion of consciousness. If we acknowledge it is only a matter of time before we discover that our galaxy is brimming with life, we can say that our galaxy is also a container dedicated to the expansion of consciousness.

Science tells us that star systems and galaxies are being born continuously, and our universe itself is expanding omni-directionally at a very high rate of speed. Therefore we can conclude that the intent of the Creative Principle, and of creation itself, is the expansion of consciousness. The Creative Principle inhabits the realm of pure consciousness that circumscribes all of physical creation. Through the Field, the Creative Principle has devised a way to connect itself to all of its expanding creation and still maintain its formless iden-tity. In other words, through the Field, the Creative Principle has omnipresence in the entire physical world without being contained by that world. The Creative Principle exists as the pure potential for all experience to occur.

The purpose of pure potential is to allow consciousness to expand by dividing itself into an infinite number of individual units of self aware consciousness that descend through layers of density, all the while becoming more and more individual. As each minute individu-al piece of the source consciousness descends into density it becomes identified with a biological form for the purposes of experiencing the sensory perception of creation, and at the same time transmitting it back to the Creative Principle.

We are, in fact, an extension of the nervous system of the Creative Principle. The collection of that sensory data is implied by the nature of all biology. We are designed to experience a continuous flow of sensory input. We collect that input and use it to manifest our reali-ty, thus becoming more aware of the nature of our existence. As this process unfolds, we come to the understanding that, as individual expressions of source consciousness, we each have the innate ability to create. We come to the understanding that we not only have the innate ability to create, but we cannot fail to use it and still exist within a biological form. The act of existing in a biological form is a creative act.

If you define a commandment as something you cannot avoid doing without ceasing to exist in physical form, there is only one real commandment imposed upon us by the Creative Principle:

"You shall create."

Whatever outer conditions we allow to be imposed upon us by our own kind, we still maintain the individual freedom to think and to feel. I define the divinely mandated requirement to act creatively in freedom as the expression of individual free will. The expression of individual free will is the method that has been selected for accomplishing the unique expansion of individual consciousness through biological forms.

If the physical universe had been set up to be inhabited by identical individual life forms all having exactly the same experience at the same time, I think you can see that condition wouldn't provide for the optimum expansion of consciousness. The most highly creative expansion of consciousness occurs through the infinite variety of individual expression. That diversity is what we see around us. That diversity is the result of the use of free will.

Religious doctrines proposing that the intent of creation is for humanity to be punished or rewarded by a higher judgmental authority, for how well we follow the rules, were all written during the dark ages of the Kali Yuga. This was a time when we thought the world was flat and the sun kept an orbit around the Earth. The Kali Yuga was a time when the intellectual abilities of humanity were at their lowest ebb.

A very large number of the same religious doctrines also propose that a higher, judgmental authority, only approves of people who commit themselves to particular closed-loop belief systems. All others are excluded from salvation by this deity who is said to be merciful and compassionate—but only if you do things the right way. As we have recently seen, some of these doctrines demand that their adherents take up arms against all who do not believe as they do.

As the quickening accelerates, more and more people are instinctively rejecting this type of philosophy for anything that is even a bit more universal. As the globalization of humanity proceeds, it is becoming clearer that the idea of religious exclusivity imposed from above does not match the evidence we've been collecting about what the intent of creation really is. If anything could be said to be mandated by the Creative Principle, diversity seems to be one of its constituent parts.

Some religious doctrines assert that every human on the planet is a sinner in need of forgiveness and salvation. This concept is a way to bind people to the idea that following the rules is the only way to be saved. It's not difficult to understand why such doctrines were written in a time when chaos and pestilence ruled the world. The intent to find a way to encourage people to live in a more harmonious manner was, and still is, a noble one. Unfortunately, 2000 years ago, we could only conceive of a relatively short-sighted belief system with which to attempt to bring about harmony.

The word "sin," as translated from the original Greek, does not mean "failure to follow the rules," it means to "miss the target." In modern parlance the word "sin" means to "miss the point." "Go and sin no more" means to "go and stop missing the point." "To be saved from sin" really means to "be saved from ignorance, from missing the entire point of one's existence," which truly is a noble endeavor.

Nearly every religious tradition on the planet tells the story of a spiritual master who transcended the bindings of ignorance and rose above the troubles of humanity. Nearly every one of those traditions say that such a master demonstrated supernatural abilities to perform what we think of as miracles. Jesus, for instance, is said to have walked on water, turned water into wine, and healed the sick. Such beings have always been present on the planet and many are with us now. They come to remind us what each and every sentient being is ultimately designed to be capable of. What do we do with those stories? We decide the master we identify with is the one and only spiritual master that has ever taken form in a physical body, and you cannot be saved unless you believe in him and him alone. If you fail to do so, you do so at your extreme peril.

Every activity we engage in has its novices, adepts, and masters. If you compete in marksmanship you find people who can barely hit the target, people who hit the target a large percentage of the time, and people who never miss. The same thing is true regarding the expansion of consciousness. There are people who are just beginning to understand the basics about human existence in the physical world. There are people who are well on their way to an understanding of human existence in the physical world and the world of spirit. Finally, there are people who have mastered all aspects of their existence completely. The latter we refer to as a "master." If you are like me, a master is someone who is very much further down the road toward a universal understanding of all creation than you are.

One of the aspects of mastery is an understanding that consciousness supercedes existence in the physical body. Even those of us who believe in a judgmental god, believe that there is some part of a human that survives the death of the physical body. After all, there must be something left to offer up for final judgment. Usually those belief systems claim that eternal life is a reward for good behavior. When asked why their judgmental god allows murder, rape, and destruction to exist, they claim that the Earth is a sinful place and the ways of God are mysterious.

Some go so far as to claim that God is locked in a battle with evil and may not win. When we accepted the premise that the Creative Principle includes and circumscribes everything in creation, we gave all that up. Since the Creative Principle encompasses everything in creation, what we call "evil" cannot exist apart from it. The darkness exists at the pleasure of the light. When you shine light into the darkness the darkness is dispelled. So what would have to be true for all the evil and mayhem that goes on here on Earth to be fine with the Creative Principle?

What would have to be true is that human consciousness is already eternal. Evil and injustice exist because they provide the ingredients for the existence of the physical plane. The duality that exists within the physical plane is what eventually turns us around and motivates us to begin making our way back into spirit.

The experiences the body is having are temporary and illusory. We have all observed that the physical body is temporary. It gets old and dies. When the physical body ceases to exist, the consciousness remains. From time to time the consciousness takes on a temporary physical form in order to participate in sensory experience. We manifest on Earth to create whatever we want, no matter how unenlightened we are. In a sense, evil is simply a form of ignorance of universal principle. Willful evil is a more concentrated version of the same ignorance. If the practitioners of evil understood that they are condemning themselves to experiencing the equal and opposite reaction of the emotional content they create around them, they might rethink what they are up to.

You can see how this process could easily be confused with the idea that the Creative Principle is doing the judging and that the eventual suffering of evil doers is the punishment that results from that judgment. However, the Creative Principle understands that consciousness is already eternal and is observing the progress you are making toward the same understanding. We must all tread the same path as our awareness develops. The higher levels of that path exist in the realms of consciousness beyond the physical world.

Even though the Bible tells us that the "kingdom of God is within you," in our culture, Christians still talk about going to heaven and going to hell, as if they were places. The emotional spiral shows that you don't go to heaven, you become heavenly. You don't go to hell, you become hellish. The kingdom of God is not a physical place; it occurs within your consciousness. It is within you. If your consciousness is heavenly, what does it matter where you are?

In Part I of this book we discussed the real nature of the physical world. We live in a universe where the very atoms that make up our world are themselves not really solid. There is in fact, no solid matter. What appears to be solid matter is really only an illusion of consciousness. Every thing we see around us is the product of consciousness interacting with the Field.

In Part II of this book I attempted to outline how we can go within and examine what is going on inside of ourselves. Mastery of what is going on inside of us leads us along the path of knowledge to a higher understanding, and eventually to the peace that passeth all understanding.

Ultimately, the injustice that appears to exist in our world is simply the product of ignorance of the true intent of creation. If we each had a more in depth understanding that the intent of creation is for all souls to experience the unlimited expansion of consciousness, we would all take much more care to see that as much as possible we do no harm to others. We would design a world around the idea of helping each other to expand consciousness rather than restricting one another from experiencing that state of joyful evolution.

Few things are as beautiful as sharing higher awareness with others. In order to do that, one must have higher awareness to share. I submit to you that the most powerful thing you can do is to heal yourself by expanding your consciousness. The second most powerful thing you can do is to inspire others to heal themselves by expanding their consciousness as well.

Chapter 28

A Cosmology for the 21st Century

As I travel around speaking to people about the intent of creation, I am often met by a variety of different kinds of blank stares. If you have spent most of this lifetime within the proverbial box, steeped in consensus reality, thinking outside of the box might seem a little weird for you. This cosmology may feel a little strange for a while, but once you get used to thinking about it in the way I have described, it gets simpler and simpler. Ultimately it is about a very simple relationship you have with the Creative Principle.

You will recall that a cosmology may be defined as "A branch of philosophy dealing with the genesis, processes and structure of the universe." A cosmology should answer the questions "Who are we?" "What is our role in creation?" and "Why do things happen the way they do?" Ultimately, our cosmology for the 21st century is a set of beliefs about the causes and consequences of our human experience here on Earth.

I freely admit this cosmology is a belief system. However, this belief system contains a method, outlined in Part II, which causes those beliefs to evolve into a personal experience, a knowing, when it is practiced diligently. That knowing, in my humble opinion, matches what is actually going on in our universe. Each and every one of us is designed to understand that our personal experience of the world and the universe we live in can be raised to one of continuous joy and eventually to bliss.

When we each remember how to do that, the need to manipulate and control others for our own satisfaction becomes obsolete. Once you learn to produce joy inside you, the pursuit of objects of sense pleasure pales by comparison. Then your focus is turned to the riches within. All of the elaborate systems we have put in place to legally confiscate the wealth of common people so a few can feel continuous sense pleasure are obsolete. All the energy that is wasted in the pursuit of continuous sense pleasure can be reinvested, bringing the standard of living of the entire population of the planet to a level as yet undreamed about. That kind of manifestation will only come if we are able to raise the consciousness of enough of humanity to cross the tipping point into a common state of higher awareness. That tipping point is closer than you might think.

In order to manifest such a state of awareness in the human population we must first believe it possible.

I began to describe our cosmology for the 21st century with a definition of the Creative Principle:

The Creative Principle is described by that set of information which not only includes but circumscribes everything in creation. The Creative Principle includes all of the atoms of all of the forms in the physical universe, yet its essential nature—being beyond form—has no form of its own.

Every concept we have considered in this book flows from this definition. By now I'm sure you understand that I have been attempting to redefine our common perception of what God is without all of the mythology. At the level of pure consciousness, the Creative Principle contains the concepts "he" and "she," but is not limited to either. The Creative Principle is an "it." Such a concept may appear to put a strain on the idea that we can have a personal relationship with our God, since all the relationships we have on Earth are with a he or a she. But I assure you that the gender-based feature of the physical world is entirely terrestrial in nature.

You already do have a personal relationship with the Creative Principle. It's in the form of a conversation. Your side of the conversation is everything you think, feel, and do in life. The response to your side of the conversation is how life itself shows up in front of you. Everything that happens to you is designed to show you the way toward fulfillment of your heart's desire. When you have created the knowing that this conversation is occurring live, in real time, all day, every day, then you'll begin to understand that your heart's desire is not an object. It's a feeling. No matter what happens in your life, what you really want, above all else, is to feel continuous joy and eventually ineffable bliss.

You may have mistaken the pleasure you experience from material objects as being the source of that joy. We are all slowly graduating from that delusion. When you have created the knowing that you are involved in a conversation with the very power that created the universe, then you round that part of your personal orbit that is farthest away from the source of all wisdom and all love, and you start to make your way back. Then you begin to know who you really are. You are the one who is in partnership with the Creative Principle itself. Together you are co-creating a dynamic, ever-new, ever-changing life. That life is not constrained by a specific plan. By its very nature, it is free to become whatever you'll have it be. Who are you? You are the co-creative partner of the Creative Principle.

The duality that is apparent in our terrestrial world is necessary in order for three dimensions to exist. If you haven't had the experience of separateness, how can you evaluate the experience of unity? The duality of good and evil, big and little, close and far is necessary not only for spiritual perspective to exist, but also for the actual visual perspective we use to navigate through the physical world to also exist. Perspective is a multidimensional concept. Creating perspective is what the Creative Principle did when it created an infinite number of self-aware units of individual consciousness. It uses us to look back at itself and know itself from a perspective other than the single point of oneness.

Why are we here? We are here because we are the embodiment of the expansion of pure consciousness, which is the same reason that the physical universe exists. The intent of pure consciousness is for the contracted consciousness of every sentient being to expand until it ultimately becomes one with the source of all. In India the pageant of creation is sometimes referred to as the divine Lila, that is to say, the divine play. The Creative Principle has designed a universe in which you can take your place as a divine playmate. The Creative Principle has allowed you the freedom to decide how it feels to participate in that play. Since it is up to you as to how it feels, why not have it feel joyous?

Why do things happen the way they do? Everything that goes on in our terrestrial theatre is the product of the creative acts of all sentient beings. The creative act of deciding how you feel about what happens, produces more of what you need to continue to feel the way you have already decided to feel. In the final analysis, the system itself is simple and elegant. You choose how you want it to feel. You develop the inner strength and knowing required for you to choose what you want independently of everything that's happening around you. That is free will. The process of exercising your free will eventually lead to wisdom. The only true wisdom is that the universe is made out of pure consciousness and that consciousness is made out of love.

No doubt I will be corrected if I have missed the point, but I believe that together we have outlined a cosmology that is worthy of our evolutionary path during the 21st century.

Chapter 29

The Planetary Free Will Experiment

Human biology has evolved as an extension of the nervous system of the Creative Principle. The Field transmits the sensory experience of every individual unit of self-aware consciousness directly to the Creative Principle. The Field is the instrument with which it simultaneously senses everything that is going on in creation without descending from its lofty perch in pure consciousness.

Native Americans, who became deeply sensitive to universal principle by living so close to nature, saw the Field as the "Spirit that moves in all things," and so it does. The Field permeates every atom in creation and not only connects us to the Great Spirit of the Creative Principle, but also deeply connects us all to each other at the subatomic level.

Everything we design seems to have an anthropomorphic quality to it. Your car mimics the human form. It has two eyes (headlights), a nose (hood ornament) and a mouth (grill). The plumbing in your home mimics your digestive system. Your computer mimics your brain. When photographed from an airplane, cars on the highway at night look just like the corpuscles that are flowing through your veins. Through our innate creativity, the human species has manifested an etheric telecommunication system that circles the globe and connects us all at higher and higher levels every day. That system mimics the Field itself.

We are electronically bombarding the planet from every direction with all the information ever generated by all humans in recorded history. The latest cutting-edge information in art, science, technology, and philosophy is getting easier and easier to acquire for all humans. This phenomenon is an evolutionary event. As I travel around and talk to people, one repetitive theme keeps popping up.

We are all connected.

It is no coincidence that seven billion of us have incarnated on the planet at this time. The age is accelerating and the quickening is at hand. The fact that there are seven billion of us now sharing the planet is encouraging all of humanity to ask: Why are we all here now? The driving force behind our growing numbers is the actual expansion of consciousness taking place throughout the universe. We are quickly approaching a quantum leap in human evolution. Whether or not we can make that leap depends on whether or not we can collectively remember who we really are.

It has been said that astral souls are lined up for the privilege of being here when humanity confronts Hamlet's question:

To be or not to be?

The growing number of people on the planet would seem to bear this notion out. Collectively we are coming to an evolutionary boundary line. On one side of that boundary line is the fear-based need we have to control, manipulate, and enslave each other. If we allow that paradigm to manifest any further, it may lead to a tragedy of overwhelming proportions yet unknown in human history. The willful restriction of human consciousness equals overspecialization. A state of overspecialization that goes on long enough, eventually leads to extinction.

On the other side of that boundary line is the possibility that we will recognize the universal success strategy of investing in the creative potential of all of humanity. Do we believe in the idea of synergy or do we not? In the matter of the expansion of consciousness, is the whole greater that the sum of its parts? If you add up all the parts of creation, is the whole of the Creative Principle still greater than the sum of the parts it has created? Yes it is. It is so by definition. There is a cosmic reason to bet on the creative potential of all humans.

Creativity equals diversity. Diversity is the hallmark of the expansion of consciousness. The expansion of consciousness leads to an understanding of the Creative Principle. Understanding the Creative Principle is the intent of creation itself and the purpose for which humanity has evolved. To that end we should develop social systems that recognize liberty as the natural state of all sentient beings and the preeminent right upon which all successful societies should be built.

The condemnation of a judgmental God is a human invention designed to bend the unaware to the will of the condemner's philosophy. I can find no evidence outside of mythology that such a thing exists in the universe. Can you? Such primitive beliefs are the basis upon which all tyranny rests. The God I have come to know is a God only of love. The power of divine love is actually what our scientists refer to as the strong magnetic force. It is the awesome force that binds one atom to another, and holds the physical universe together.

The use of fear to motivate people to any end is not a sustainable strategy for the continuation of the human species in the higher ages. The quickening is manifesting the essence lying behind our belief systems. We are surrounded by the process of getting what we are, not what we want. That is why the world is in such a state. In order to get what we want, we will need to become what we want. It is time to make a unified effort to upgrade our belief systems to a cosmology that is worthy of our highest ideals in the 21st century.

You may not agree with all that has been presented in this book. However, I will stand for your right to believe anything you want so long as you do agree to do no harm to others. John Stuart Mill, considered one of the most influential English-speaking philosophers of the 19th century, said:

"Liberty resides in the rights of that person whose views you find most odious."

We all have a right to decide what to believe. If we can establish a universal commitment to do no harm to others over what we or they believe, then the world will be a much more peaceful place. If you wish to do harm to yourself with your philosophy, then on your own head be it. Mill went on to say:

"The only purpose for which power can be rightfully exercised over any member of a civilized community, against his will, is to prevent harm to others. His own good, either physical or moral, is not a sufficient warrant. . . Over himself, over his body and mind, the individual is sovereign."

We need to recognize the sovereignty of the individual with renewed enthusiasm. We are not sheep, nor cattle. We are each a holy subdivided part of the Creative Principle itself. Our innate potential is to live lives of continuous inspiration in the joy of realizing who we are and why we are here.

We are involved in a planetary free will experiment. It is planetary because it involves all of us. It is an "experiment" because the outcome is not assured. What will make the difference is how we choose to use our inalienable right to express the free will to create. You were born with the power to call forth your own reality. In fact, you've been calling it forth your whole life. It is now time for you, and the rest of humanity, to use that understanding to rise to the next level of human evolution.

It has been my great pleasure and privilege to have organized the mostly ancient information in this book for you. The not-so-unique experience that qualifies me to have done so is that I've been to the bottom. I know the halls of hopeless despair. Having slowly made my way back up the emotional spiral, I finally understand that the price of joy is eternal vigilance in the here and now. I look forward to our discussions in the near future.

By now, I sincerely hope you recognize the immense power of the Creative Principle that resides within you. I'd like to leave you with one final question:

What will you decide to create next?

William O. Joseph
312 Dwapara

Notes

Chapter 4: Inner Space

1. John Campbell, Rutherford Scientist Supreme (Washington, D.C.: American Astronomical Society Publications, 1999)

2. For an easy to understand animation of the double slit experiment see: YouTube.com/watch?v=Q1YqgPAtzho.

Chapter 5: The Field

1. Swami Sri Yukteswar Giri, The Holy Science (Los Angeles, CA.: Self-Realization Fellowship, 1977): p. 8.

Chapter 6: The Law of Attraction

1. Paramahansa Yogananda, The Autobiography of a Yogi (Los Angeles, CA.: Self-Realization Fellowship, 1946): pp. 109–141.

Chapter 7: A Quantum Leap

1. Paramahansa Yogananda, The Autobiography of a Yogi (Los Angeles, CA.: Self-Realization Fellowship, 1946): p. 417.

Chapter 9: Co-creation and the Law of Unintended Consequences

1. Robert K. Merton, "The unintended consequences of purposive social action.," American Sociological Review, vol. 1, no. 6 (December 1936): pp. 894–904.

Chapter 12: The Precession of the Equinoxes

1. See NASA's explanation at Www-spof.gsfc.nasa.gov/stargaze/ Skepl2nd.htm.

2. Paramahansa Yogananda, The Autobiography of a Yogi (Los Angeles, CA.: Self-Realization Fellowship, 1946): pp. 420–40.

Chapter 18: The Spell Is Cast by Paper Money

1. For a history of Sweden's Central Bank, visit the Sveriges Riksbank website: Riksbank.se/en/The-Riksbank/Historia1/Money-and-power-the-history-of-Sveriges-Riksbank/Stockholms-Banco.

2. I first heard the term "Grey Men" used in 1988 by J.Z. Knight to describe those who manipulate the money system.

Chapter 22: The Long Con in American History

1. G. Edward Griffin, The Creature from Jekyll Island (Westlake Village, CA.: American Media, 1994): p. 5.2. Frank A. Vanderlip as cited by Name of Reporter, "From Farmboy to Financier, "Saturday Evening Post (February 9, 1933): pp. 25-70.

Recommended Reading

Paramahansa Yogananda. *The Autobiography of a Yogi.* Los Angeles, CA.: Self-Realization Fellowship, 1946.

Swami Sri Yukteswar Giri. *The Holy Science.* Los Angeles, CA.: Self-Realization Fellowship, 194977.

R. Buckminster Fuller. *Critical Path.* New York: St. Martin's Press, 1981.

R. Buckminster Fuller. *Operating Manual for Spaceship Earth.* New York: E.P. Dutton, 1978.

R. Buckminster Fuller. *Grunch of Giants.* New York: E.P. Dutton, 1983.

James Gleick. *Chaos.* New York: Penguin, 1987.

G. Edward Griffin. *The Creature from Jekyll Island.* Westlake Village, CA.: American Media, 1994.

Amit Goswami, Ph.D. *The Self-Aware Universe.* New York: Putnam, 1995.

Neale Donald Walsch. *Conversations with God.* New York: Putnam, 1996.

Graham Hancock. *Underworld, the Mysterious Origins of Civilization.* Toronto: Doubleday Canada, 2003

About the Author

William O. Joseph grew up in a small town in Ohio and then was drawn to the American West in his early 20s. He spent many years living close to nature while exploring the Rocky Mountains on horseback. He enjoyed a 40-year career in architectural design, construction, and real estate development. He has lived in New Mexico, Colorado, Arizona, California, and Idaho, and now lives and writes in Montana.

4152434R00164

Printed in Great Britain
by Amazon.co.uk, Ltd.,
Marston Gate.